A BASIC TREATISE ON THE COLOR SYSTEM OF WILHELM OSTWALD

THE COLOR PRIMER

EDITED AND WITH A FOREWORD AND EVALUATION BY FABER BIRREN

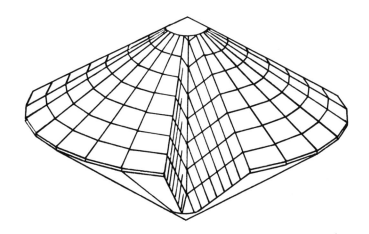

 VAN NOSTRAND REINHOLD COMPANY
New York Cincinnati London Toronto Melbourne

Van Nostrand Reinhold Company Regional Offices: New York, Cincinnati, Chicago, Millbrae, Dallas
Van Nostrand Reinhold Company International Offices: London, Toronto, Melbourne

Copyright © 1969 by Litton Educational Publishing, Inc.
Library of Congress Catalog Card Number 69-15897
ISBN 0-442-11344-7

All rights reserved. No part of this work covered by the copyrights hereon may be reproduced or used in any form or by any means—graphic, electronic, or mechanical, including photocopying, recording, taping, or information storage and retrieval systems—without written permission of the publisher. Manufactured in the United States of America.

Type set by Lettick Typografic Company, Inc.
Printed by Halliday Lithograph Corporation
Color printed by Princeton Polychrome Press
Bound by Publishers Book Bindery, Inc.

Published by Van Nostrand Reinhold Company
450 West 33rd Street, New York, N.Y. 10001
Published simultaneously in Canada by
    Van Nostrand Reinhold, Ltd.

3 5 7 9 11 13 15 16 14 12 10 8 6 4 2

Contents

FOREWORD	5
WILHELM OSTWALD	7
HISTORY OF COLOR SYSTEMS	9
THE COLOR PRIMER	17
Introduction	
First Chapter, The Achromatic Colors	19
Second Chapter, The Chromatic Colors	30
Third Chapter, Light Clear and Dark Clear Colors	40
Fourth Chapter, The Muted Colors	49
Fifth Chapter, The Color Solid	58
Sixth Chapter, The Harmony of Colors	65
AN EVALUATION	72
REFERENCES	79
COLOR PLATES	81

Foreword

The Color Primer of Wilhelm Ostwald (*Die Farbenfibel*), translated by permission of Akademie-Verlag and reprinted herewith, is one of the most successful books ever published on the subject of color. While Ostwald's larger work, *Colour Science* (in two volumes), went into English in 1931 and was issued by Winsor and Newton in London, the more elementary *Primer* was not so honored until now.

Ostwald in his day was something of a German institution. In fact, during the latter years of his life the sum of his accomplishments, his books, charts, solids, and the like were combined in a commercial organization known as "Ostwald Energie," and this organization not only reached the entire German educational system but offered to extend counsel to German business and industry.

Being a Nobel prize winner and a foremost scientist, Ostwald had the backing and support of the German government, and his theories and principles became more or less mandatory as training aids in German schools. Even when Hitler came to power in 1933, a year *after* Ostwald's death, the new government continued to advance Ostwald's ideas and to see to it that Germans, young and old, properly respected the scientist's views.

Yet while Ostwald's eminence was highly regarded in academic and business circles, the field of art was not so gracious. There is little evidence that the famous school of Expressionism in painting, which rose in Germany and the Nordic countries during Ostwald's life, had much interest in the chemist's theories.

At the German Bauhaus, however, founded in 1919 and devoted to highly original and remarkable developments in architecture, painting, sculpture, industrial design, the graphic arts, and decorative arts, Ostwald was far from unknown. Walter Gropius, the founder, installed Ostwald's charts and spinning devices in the lobby of the school. Great Bauhaus teachers, such as Josef Albers, Oskar Schlemmer, Laszlo Moholy-Nagy, all knew of Ostwald, held major or minor regard for his work, and made small or large use of his books and charts. Some actually met him. Among those who held personal discussions were two of the outstanding abstract painters of modern times, Wassily Kandinsky and Paul Klee.

Hitler closed down the Bauhaus in 1933, called much of its work degenerate and confiscated many of Kandinsky's canvases. Yet the prestige of Wilhelm Ostwald lived on into a posterity that is still very much alive and animate.

The Germans, of course, still revere Ostwald as a color theorist. In England, Ostwald's system and ideas of harmony are widely sponsored and taught in British schools. Other writers and scholars have carried on the tradition of Ostwald's genius.

In America, while some more advanced schools in the field of higher education teach of Ostwald, the American Albert H. Munsell and the venerable red-yellow-blue theory continue to predominate. However, in the commercial world, Ostwald's system as produced by the Container Corporation of America has been given widespread application. Except for color notation, where Munsell has major acceptance, the Ostwald solid in the form of the *Color Harmony Manual* (see References at end of book) offers probably the most highly regarded series of color standards in American use today.

Finally, because of the exceptionally harmonious color relationships which are inherent in Ostwald's solid, many outstanding and creative artists have gone to it as a wellspring of color beauty and magnificent visual order.

FABER BIRREN

Wilhelm Ostwald
(1853-1932)

Wilhelm Ostwald was born in Riga, Latvia, on September 2, 1853. His parents, however, were German and favored their son with a liberal as well as scientific education. After private and academy schooling he attended the University of Dorpat in eastern Estonia, where he showed remarkable intelligence and competence.

In 1882, at the age of twenty-nine, he became a professor of science in Riga. Being a true scholar and linguist (he spoke English and French as well as German and the local Russian dialect of Latvia) his abilities became widely known. He had an alert mind and held universal interests in the sciences and the arts. And he was tall, genial, and handsome.

In 1887, at the age of thirty-four, he was appointed professor of physical chemistry at the University of Leipzig. Here he remained for nearly twenty years, eventually becoming director of the Physio-Chemical Institute.

Having the qualities of genius, Ostwald had an original mind, plus great powers of concentration, and an unusual capacity for hard work. In addition to his duties as a teacher, he did creative work in his chemical laboratory, was an ardent biographer of scientific men, the author of several highly important works on chemistry, and the founder and editor of scientific publications.

As a chemist Ostwald won world renown, his foremost accomplishments being in the field of electrochemistry and chemical solutions. For his discovery of a method of oxidizing ammonia to form oxides of nitrogen (among other things) he was awarded the Nobel Prize for Chemistry in 1909.

However, Ostwald had retired from the University of Leipzig in 1906, at the age of fifty-three, to live in Saxony. Here he began a new life, devoting himself to natural philosophy, to technical problems in the art of painting, *and to color*!

What Ostwald achieved as to color theory and color organization is the substance of this book. Having a broad and philosophical viewpoint, he was able to convert a wealth of technical knowledge to artistic ends. On art, for example, he wrote *Letters to a Painter on the Theory and Practice of Painting* which was published in English in 1907. In this work he very capably dealt with driers used for artists' paints, with the grinding of pigments, the making of pastel crayons, and with the protection of canvas.

On color, he wrote numerous works, the first of which was published in 1916. Here his fame was instantaneous. His *Farbenfibel* (Color Primer) ran into fifteen editions. Ostwald colors for the duplication of the Ostwald System became available in powdered tempera, dyed wool, and dyed papers. In a catalog issued by his publisher as late as 1930 over 20 items, books, color solids, color atlases, scales, charts, and coloring material, were offered, all the creations of Wilhelm Ostwald.

Ostwald came to America in 1905, lectured at the Massachusetts Institute of Technology and was honored across America by the scientific and educational world. He met Albert H. Munsell at this time, visited Munsell's studio and had several discussions about the American's theories and color sphere. However, Ostwald was yet to create and design his own triangle and solid. What is significant, perhaps, is that the two men responsible for the two most important color systems of this century had known each other.

History of Color Systems

This chapter has been written by Faber Birren as an introduction to color systems and color solids at large and to the Ostwald system in particular. The world of color is a wide and varied one. In visual sensation are thousands of hues, tints, shades, and tones which are seen everywhere in natural and manmade objects and environments. Although a certain amount of order will be found to exist in a rainbow or a sunset, for the most part nature uses color in a lavish and indifferent way. How is one to take all that is seen by the eye, organize it, sort it out, and arrange it in a sensible and tidy manner?

When Sir Isaac Newton declared that all colors were to be found in white light (around 1666) he designed the first of all color circles, drawing together the two ends of the spectrum, red and violet, and spacing orange, yellow, green, blue, and indigo between them. (See a companion volume in this series, Faber Birren, *Principles of Color*.) Newton was concerned here with pure spectral colors, and he went no farther. Yet the eye also sees colors that are composed of hues mixed with white, with black, and with gray. If there are relatively few clearly distinguishable colors in a pure spectrum, the white-containing, black-containing, and gray-containing colors are profuse indeed.

Any chart, diagram, system, or solid that hopes to plot the world of color in an orderly way must be conceived in terms of three dimensions. According to Ogden Rood, a distinguished American scientist in the field of vision and color (1831-1902), a first attempt to devise a color solid was made by a German mathematician, Tobias Mayer, in 1758. Mayer arranged a series of triangles as interpreted in Figure 1. A key triangle contained pure red, yellow, and blue on its angles. These colors were then mixed to form secondary colors along the edges of the triangle, and dull tertiary tones toward the center. Other triangles with white additions were placed above the key triangle, and still

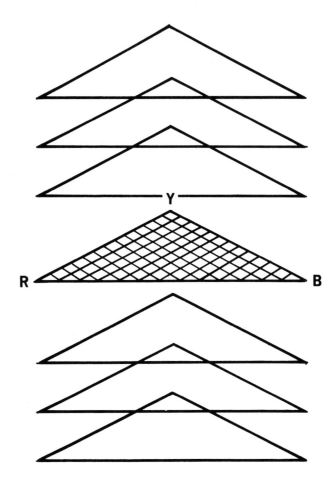

Figure 1. The color solid of Tobias Mayer, 1758.

others with black additions were placed below. Unfortunately, dull mixtures which occurred in the key triangle were repeated in the other triangles. To all indications Tobias Mayer never found this out, for his work was never published and his solid never built.

In 1772, J. H. Lambert, an English physicist and philosopher, did better.

(See Figure 2.) He also began with a triangle having red, yellow, and blue on its angles. He noted — correctly — that mixtures of pigments were subtractive. Hence the base triangle had black for its center. Above this triangle Lambert arranged progressively smaller triangles scaling toward a white apex. Here again he wisely observed that colors became less distinguishable, and therefore fewer in number, as white was approached.

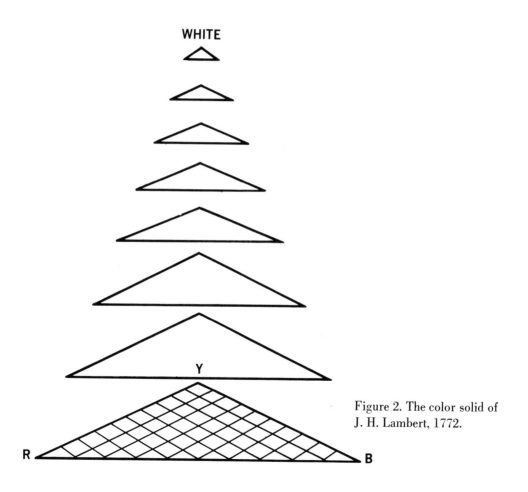

Figure 2. The color solid of J. H. Lambert, 1772.

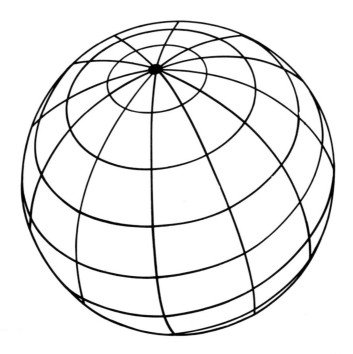

Figure 3. The color sphere of Philip Otto Runge, 1810. Base colors red, yellow, blue.

In 1810, an important German classical painter by the name of Philip Otto Runge designed the color sphere shown in Figure 3. This was illustrated in a book called *Die Farbenkugel* (Color Globe). Curiously, Runge's treatise was issued in the same year that a master work on color by the great German poet Goethe was published, *Farbenlehre* (Doctrine of Colors). In Runge's sphere, pure colors ran about an equator and had gradations of a primary red, yellow, and blue. These pure colors then scaled up toward a white apex, down toward black, and in toward gray. In this conception Runge came close to an ideal solid in which all distinguishable colors could be included. Albert H. Munsell of America followed the same principle, with due credit to Runge, and so did

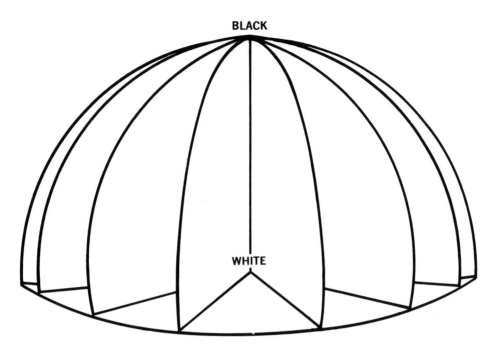

Figure 4. The color hemisphere of M. E. Chevreul, 1839.

other later theorists. Runge, who was born in 1777, died in 1810 — the year in which his book was issued — at the young age of 33.

After Runge, other but less significant solids were designed — up to the remarkable inventions of Albert H. Munsell and Wilhelm Ostwald. In 1839, M. E. Chevreul of France conceived of a color hemisphere interpreted in Figure 4. Although Chevreul was one of the most influential colorists of all time, and although his masterwork, *The Principles of Harmony and Contrast of Colors*, influenced the schools of Impressionism and Neo-Impressionism, his color solid was both inadequate and impractical, and a model of it was never completed.

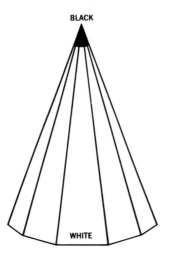
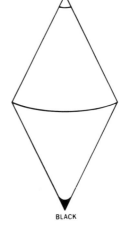

Figure 5. The color cone of Wilhelm von Bezold, 1876.

Figure 6. The double cone of Ogden Rood, 1879.

In 1876, Wilhelm von Bezold of Germany, following a pattern set by the eminent physicist and physiologist Hermann von Helmholtz, devised the pyramid or cone interpreted in Figure 5. Choosing additive primaries for the base of the cone, these scaled toward a white center. Lambert's idea was reversed in that gradations were included — and diminished in number — toward a *black* apex.

In 1879, Ogden Rood of America devised a color solid interpreted in Figure 6. This had pure colors around the circumference which scaled up toward white, down toward black, and in toward gray. The principle was similar to that of Runge. Wilhelm Ostwald, who followed this pattern, declared, "The Double Cone must be regarded as the final and enduring solution to the color solid problem." This confirmed what Rood had previously stated: "In this double cone, then, we are at last able to include all the colors which under any circumstances we are able to perceive."

There are three further color solids to mention before treating with the Ostwald System in detail. The first of these is the Munsell solid shown in Figure 7. However, no description will be given here: the reader is referred to a companion volume in this series, Albert H. Munsell, *A Grammar of Color*.

In Figure 8 is the color pyramid of A. Höfler. This is meant to serve as a three-dimensional diagram of the way in which color sensations find order and relationship in human experience. Höfler's Pyramid is widely referred to in the field of psychology and will be found in most textbooks on this science. First is a neutral gray axis scaling from white (W) to gray (GR) to black (B). On the four corners of the central section are the four primary colors in vision, red, yellow, green, blue. All color variations and gradations lie within the figure, whitish tints scaling up toward white, blackish shades scaling down toward black, and grayish tones scaling in toward the gray axis.

Figure 7. The color solid of Albert H. Munsell, 1900.

Figure 8. The color pyramid of A. Höfler, 1897.

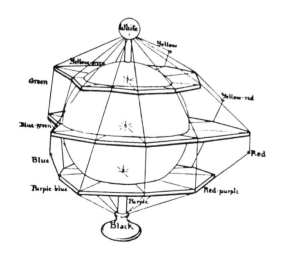
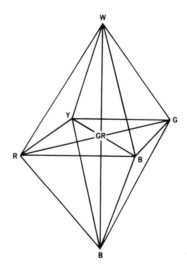

15

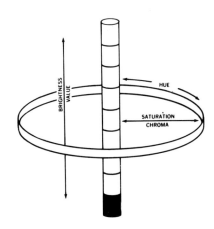

Figure 9. Diagram of three dimensions for color: hue, brightness (value), and saturation (chroma).

In Figure 9 is a similar diagram (to Höfler) which is commonly used today to illustrate simple color dimensions. The Ostwald System tends to follow the order of Höfler's Pyramid (Figure 8), while the Munsell System tends to follow Figure 9.

In accordance with Figure 9, the dimension of hue can be indicated on a belt or circumference around which pure colors are placed.

Rising up and down is a central pole of neutral grays scaling from white to black. The dimension of brightness or value (Munsell) is found in colors that scale vertically, and which have degrees of brightness corresponding to those of the gray scale.

A third dimension of saturation or chroma (Munsell) is found in colors that scale horizontally from the outer perimeter of pure hues to the inner core of neutral gray.

The facts of color sensation are well expressed both in Figures 8 and 9 and both hold true to average experience. It is now the pleasant task of this book to feature the remarkable work of Wilhelm Ostwald, to explain in simple terms the novel design of his system and to illustrate how remarkable principles of color harmony may be drawn from it.

The Color Primer

By Wilhelm Ostwald

(*Note:* Here follows a complete, liberally translated and edited reprinting of Wilhelm Ostwald's *Die Farbenfibel*. This reproduces more or less intact all the text and all the black-and-white and color illustrations of the German edition. Few liberties have been taken other than to use English color terminology familiar to English readers and recognized in the literature of English books on color.)

INTRODUCTION

Color. Everything we see consists directly of colors which are spread out in the field of vision as larger and smaller parts or areas. Where two or more areas meet, borderlines are created, the continuity of which brings about the forms or figures from which we sense the presence of objects seen.

The colors are, therefore, the basic components or elements of our sensation of vision.

Because of inexact use of language, the word color is also being used for materials and processes through which colored sensations or colors are created. Materials serving for dyeing are called dyestuffs and not colors; the energy which creates the sensation of color by acting upon our eye is called light.

Division. Colors fall into two classes:
1. The achromatic colors white, gray, black, and everything between them;
2. The chromatic colors yellow, red, blue, green, and all that are between and adjacent to them.

At times, the designation color is restricted to chromatic colors and the achromatic ones are called colorless. But then one could no longer say that all sensations of vision consist of colors. It is more appropriate, therefore, to select the wider linguistic usage (that white, gray, and black also are colors).

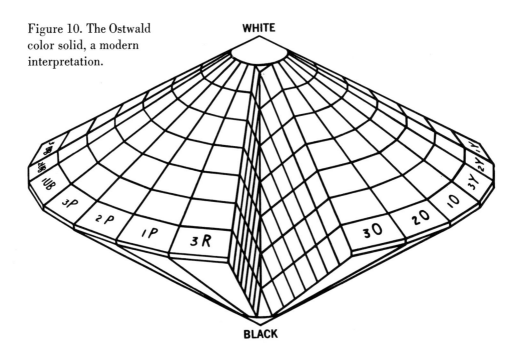

Figure 10. The Ostwald color solid, a modern interpretation.

FIRST CHAPTER
The Achromatic Colors

The Order. The achromatic colors form a constant simple (one dimensional) series with end-points in black and white. Between them, all gray colors can be arranged in such a manner that each will receive a definite, determined place between its neighbors. One neighbor is lighter, the other darker than the given gray.

If, for example, 5 gray colors 1 to 5 are given in random order:

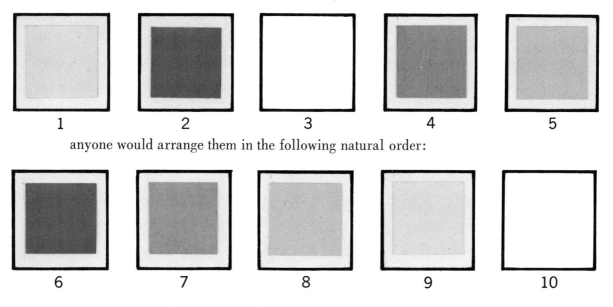

anyone would arrange them in the following natural order:

and there would be no person whose sensation of color would demand another order. In this series, 6 is the darkest, 10 the lightest color. These designations express the fact that of the light falling onto this paper, surface 6 reflects the least and 10 the most. For each link or step in this series, all steps to the right of it are lighter, all those to the left, darker. For example, if 8 is lighter than 7, and 7 is lighter than 6, 8 will also be lighter than 6 (and so forth).

If another gray is brought in that is not yet present in the series, then there will be only one place where it belongs. Thus, gray 11

11

can be arranged only between 7 and 8 in such a manner that it will be lighter than those steps to the left of it and darker than those to the right. But if it is introduced, for example, between 8 and 9 then all grays to the right will be lighter, but not all those to the left will be darker, inasmuch as 8 is lighter than 11.

The gray series is thus specific.

Continuity. Between two different grays it is always possible to insert a third gray, which is lighter than one and darker than the other. In this manner the steps can be made even smaller, until they finally become imperceptible.

The complete gray series is thus continuous as in 12.

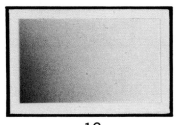
12

It probably follows that the complete gray series consists of an infinite number of steps. However, if one places between two terminal points a series of gray sheets, each of which is just noticeably lighter than the previous one, it will be found that one cannot discern an infinite number of intermediate steps. Rather, a finite difference is necessary if one is to distinguish a series of grays, and if the steps become very small, differences can no longer be discerned.

The Threshold. This border between just noticeable difference in color is called the threshold.

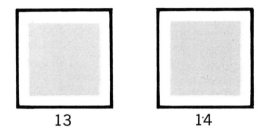
13   14

Thus, there is between 13 and 14 a just noticeable difference, while between 15 and 16 there is an unnoticeable difference. Even though 16 is a trifle weaker than 15, they both appear equally as light.

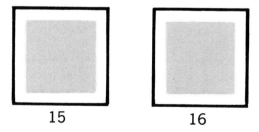
15   16

Equality. Only the presence of the threshold makes it possible for us to regard two gray colors as equal. What we cannot distinguish we call equal. Even if we could recognize every difference that actually existed (in gradations) it would be impossible to create two equal grays, as we could never remove the last traces of the actually existing differences. In effect, we will regard two gray colors such as 15 and 16 as equal, even though an objective difference *between them* has intentionally been created.

The presence of the threshold has certain consequences with regard to the apparent equality of colors. These consequences are different from the mathematical relationships that are usually established when no regard is paid to the threshold. For example, in general this law applies: if $a = b$ and $b = c$, then $a = c$. And it also follows from $a = b$, $b = c$, $c = d$, that $a = d$. Now, if gray b is indeed lighter than a, but by less than the threshold, and if the same applies between c and b, and between d and c, we would first state $a = b$, $b = c$, and $c = d$. But if the sum of these imperceptible differences exceeds the limits of the normal threshold, then we could by no means say that $a = d$, but we would experience d as lighter than a.

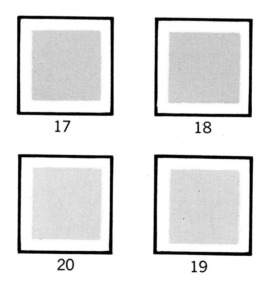

17   18

20   19

Thus, there exists a difference between steps 17 and 18 that is smaller than the threshold. Therefore, if steps 19 and 20 are covered, 17 and 18 will appear equal. Similarly, 18 and 19 appear equal if 17 and 20 are covered, and the same applies to 19 and 20. We have observed, therefore: $17 = 18$, $18 = 19$, and $19 = 20$ and are thus inclined to conclude also that $20 = 17$. However, if we cover 18 and 19 and compare 20 and 17, step 20 is unmistakably lighter than 17.

Incidentally, the threshold is not an invariable value, as it often has a different value with different persons. Much depends on visual ability; in some individuals the threshold may increase or shrink through exercise or through fatigue and other weakening influences. For this reason, some will not experience the difference between steps 13 and 14, while others will notice a difference between 15 and 16. The number of distinguishable steps of gray under normal conditions amounts to several hundred.

Brightness. Reasons for different sensations of gray are found in the differing intensities of reflected light. This does not, however, apply to absolute luminous intensity, for a medium-gray paper lying between black and white will retain its color, whether it is illuminated with a strong or with a weak light. Rather, what governs is the amount of light falling on a surface that is reflected back. We call this BRIGHTNESS, because the color of a surface will be brighter as it receives more light. If *all* the light is reflected (whereby it is dispersed in all directions and not mirrored in certain directions) the surface is called full *white*; if all is absorbed and none is reflected, it is called full *black*. If a part is reflected, it is called *gray*.

These rules apply to those surfaces which *uniformly* reflect or absorb *all types* of light. However, if the area acts selectively, so that certain types of light are reflected more amply than are others, then its color will no longer be white or gray, but chromatic.

White. The best approximation of perfect white is considered to be a dull coat of pure barium sulphate. We put its luminosity at 100. Best zinc white (21) has a luminosity of 92, i.e., a coat of zinc white reflects 92/100 of the amount of light falling on it. Chalk white (22) has a luminosity of only 80 and in addition is also yellowish, i.e., chromatic. (NOTE: the two illustrations show the paper surface only and thus do not accurately convey Ostwald's original samples. Zinc white as a pigment is fairly clear, while chalk white is slightly yellowish.)

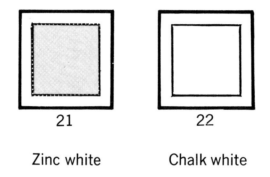

21   22

Zinc white   Chalk white

Designation. Each gray is designated by the fraction of light it reflects, expressed in percentages. Thus, a gray 25 reflects 25/100 or 1/4 of the light, and a gray 4 or 04 only reflects 4/100 or 1/25. Such a gray would still be called black. A gray 80, which reflects 80/100 and absorbs 20/100 or 1/5 of light, would still be called white.

Black. A black that will not reflect any light can be produced by making a 4-inch cube-shaped box of dull black painted cardboard, with the black surface facing inside, then cutting an opening about 3/4-inch-square in the center of one side. This opening will be much blacker than any black coat of paint, for the latter will appear gray if held beside it. Even the best black silk velvet, the blackest there is, will show in this manner that it still reflects appreciable quantities of light. Good black dyestuffs, dully applied, usually have a reflectance of 02 or more.

Stages of Gray. If one wishes to insert between light and dark gray (or white and black) an intermediate step in the middle, the nearest thing at hand is to select for this an achromatic color of medium brightness. Thus, gray step 23 has a brightness of 80, 25 has a brightness of 5, and 24 has the mean value of $\frac{80+5}{2} = 42.5$. However, no one will admit that gray step 24 is in the center between 23 and 25. Rather 24 will be seen as being too light.

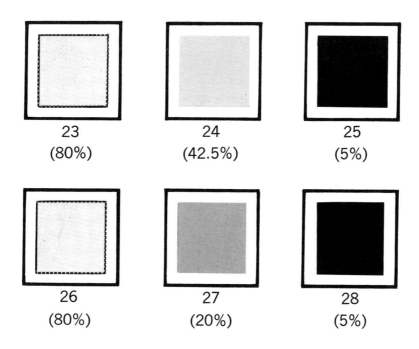

On the other hand, in gray steps 26, 27, and 28, step 27 will be recognized as being in the center. Yet gray step 27 has a brightness of 20; thus, there is a brightness difference of 60 against step 26, and of only 15 against step 28, i.e., a quarter of it.

On the other hand, there is the following relationship: 80:20 = 4 and also 20:5 = 4. The series having 5, 20, 80 brightness is thus a "*geometrical* progression," i.e., one whose components increase in difference but at the same rate from step to step. Against this, 5, 42.5, 80 was an "*arithmetical* progression," whose components increase by the same amounts from step to step.

Thus, where brightness differences form a geometrical progression, only then do we experience corresponding gray colors as being visually equidistant.

The Gray Scale. Of the innumerable geometric progressions according to which the steps of gray can be arranged, there shall henceforth be used only that in which ten steps each are inserted between 1 and 10 and between 10 and 100. The figures can be shown exactly only by infinite fractions; because of the threshold, however, such exactness is of no importance and rounding off to two digits is reasonable. In descending order the series (progression) will read:

| 100 | 79 | 63 | 50 | 40 | 32 | 25 | 20 | 16 | 12.6 |
| 10 | 7.9 | 6.3 | 5.0 | 4.0 | 3.2 | 2.5 | 2.0 | 1.6 | 1.26 |

The last components of this series rarely occur; they are used to establish the standards of the achromatic colors, i.e., those colors which are used to the exclusion of all others wherever one has free choice. For this purpose, one has to take in all instances the mean values between two successive steps. In this manner the norms of gray have been established.

89	71	56	45	36	28	22	18	14	11	8.9	7.1	5.6	4.5
a	b	c	d	e	f	g	h	i	k	l	m	n	o

3.6	2.8	2.2	1.8	1.4	1.1	0.89	0.71	0.56	0.45	0.36
p	q	r	s	t	u	v	w	x	y	z

There is a letter under each norm, by which the latter will henceforth be designated. First, each letter represents the corresponding gray; e.g., **h** is the gray with 18 percent white and 82 percent black. Later, the same letters will be used to designate the corresponding amount of white or of black alone; thus **h** could also mean 18 white or 82 black.

In practice, these steps have proven to be too narrow. Every other one is therefore skipped, and thus the practical gray scale is obtained —

89	56	36	22	14	8.9	5.6
a	c	e	g	i	l	n

3.6	2.2	1.4	0.89	0.56	0.36
p	r	t	v	x	z

The gray scale (29-36) is shown on a separate page. Its 8 steps run from **a** down to **p**. A deeper black than **p** could not be created by normal printing processes.

A thorough examination of this gray scale shows that the distances are indeed experienced as being equally spaced to the eye. Only the lower steps give a more crowded impression, in view of the circumstance that the law of geometrical progression no longer exactly represents the facts among dark colors approaching black.

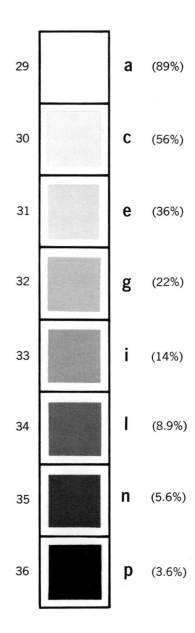

Figure 11. The gray scale. Figures in percentages indicate white content. Black content is the sum that remains to equal 100.

Consequences. From the law of the geometrical progression there results a contrasting behavior of the white and the black ends of the achromatic series. While large amounts of black cause only faint changes among light grays — or only slightly influence the appearance of the gray colors — the slightest quantities of white added to black among dark grays will cause a very marked lightening, and this becomes even more marked with rich, pure black. Thus it can be seen from the brightness series just described and separately illustrated that the addition of 33 per cent of black at the white end (the difference between 89 and 56) will not cause a greater difference than 2 per cent of white added to the black end (the difference between 3.6 and 5.6).

It is therefore necessary for all artistic and commercial purposes to arrange the gray scale not according to the numerical values of the content of white, but according to the same steps of sensation as has been done here. If one were to arrange them according to equal (arithmetic) steps of contents of white, one would obtain too many steps at the white end and too few at the black end.

Norms. The just designated achromatic steps a c e g i l n p r t etc. shall henceforth be regarded as NORMS. This means: if one enjoys sufficiently free choice (and this is almost always the case), one does not use just any assortment of grays, but rather one or more out of the series a c e g i l n p r t v x z which are in orderly arrangement. From this, very great advantages are to be derived as to achromatic color harmonies.

SECOND CHAPTER

The Chromatic Colors

(Editor's note. All color illustrations described in the pages that follow will be found at the end of this book on Plates I through VIII. Ostwald's numbering system, 37, 38, 39, 40, etc., has been preserved throughout so that text references will agree with what is shown on the Color Plates.)

The Diversity. A given achromatic or gray color can be changed in only one way: it can only be made lighter or darker. The achromatic colors therefore form a one-dimensional group.

On the other hand, every chromatic color can be changed in several ways. (See Figure 12.) One can shift a given color in the following manner:

a) One can make a red more yellowish or more bluish, a blue more reddish or greenish, a green more bluish or more yellowish. This is called changing the *hue*. This is shown in the colors from 37 to 40 on Color Plate I. The sequence is from orange to red.

b) One may retain the hue and add increasing fractions of white to the pure color. The variations will thus become increasingly lighter. The colors 41 to 44 on Color Plate I have the same hue, but contain increasing amounts of white. The scale runs from purple to orchid.

c) One can darken the hue by adding increasing quantities of black. The colors 45 to 48 on Color Plate I have the same hue with increasing amounts of black. The scale runs from orange to brown.

d) One may add *both* white and black to the pure color; this is equivalent to adding a certain amount of gray of corresponding brightness. The colors 49 to 52 on Color Plate I have the same seagreen hue to which increasing amounts of a medium gray have been added.

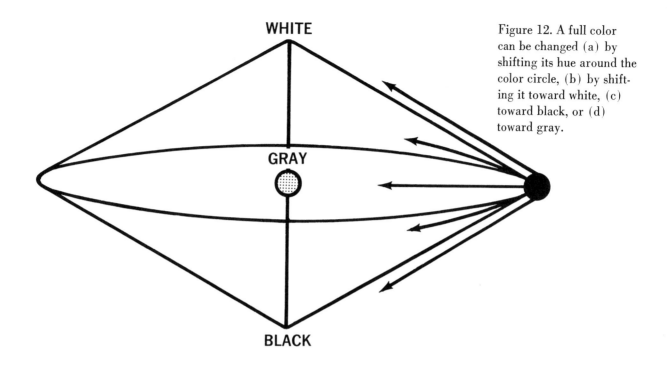

Figure 12. A full color can be changed (a) by shifting its hue around the color circle, (b) by shifting it toward white, (c) toward black, or (d) toward gray.

Case d may be taken as a generalization or combination of cases b and c, as white and black are the terminal points of the gray series and combine to form gray.

Therefore, any color may be regarded as consisting of a pure color of given hue, white, and black. All three together make up the color.

There exist no further elements than pure hue, white, and black to form all colors, achromatic or chromatic. The chromatic colors form a three-dimensional group, as will be seen.

The Hue. The color designations red, green, yellow, blue, etc. refer to that characteristic of the chromatic colors which is called the hue. The chromatic colors differ from the achromatic colors by reason of the presence of the hue.

A color that expresses only a definite hue and has no achromatic component is called a full color (saturated color). Actually, most average colors such as seen in painted, dyed, or natural materials contain, in addition to the full color, also an achromatic portion, namely white and black.

The Hue Series. As with the brightness steps of the achromatic colors, hues also form a constant series. Given any miscellaneous collection of colors, they will be arranged in related order by various observers. Some will experience certain hues as being closer or more similar, and others as being more distant and less similar. A random series of chromatic colors, such as shown in 53 to 60 will be arranged in the sequence from 61 to 68 if one starts with yellow and continues to red. (See Color Plate I.)

However, while no exceptions could be observed in the achromatic luminosity series, there exist certain individual persons (usually color blind) whose judgment concerning chromatic colors is incomplete or even deviating. We shall disregard these exceptions and shall restrict ourselves to people with similar judgment; they form the great majority, the "color-efficient" ones.

The Hue Circle. Unlike the achromatic series, which has definite terminal points in white and black, the hue series has no such distinguishing points. Rather, we could start the series with any desired hue and would always find the following. First, the subsequent hues will become increasingly more remote from the first hue. But this goes on only up to a certain spot, and from there on the hues of the initial color become again increasingly more similar, so that the last will be most similar to the first. Thus, the series of hues turns around and forms a ring which can most simply be represented by a circle. Instead of arranging the hues along a straight line, as in 61 to 68, it is better to arrange them in a circle, such as in 69 to 76. (See Color Plate II.) Here, the hue series can be started at any point and will return to this point, once it has run through the entire series.

Starting Point and Progressive Steps. To be able to designate the individual links of the series one must start at a point chosen arbitrarily. We choose as a starting point of the hue series the lightest point, a pure yellow that is neither greenish nor reddish.

In addition, the progressive hue steps of the series must also be fixed arbitrarily. In the gray series there exists a single-directional difference between darker and lighter. However, such a progressive difference does not exist in the hue circle; from the yellow point one could just as well turn to the red as to the green side. We have decided that from yellow we shall always advance towards the red side.

Continuity and Threshold. Just as with the gray series, the hue series is continuous. Between two different hues it is always possible to insert a third one that will be more similar to the two than the two to each other, and this process can be continued until no differences can be noted. Then the hues are equal.

Here too, there exists a threshold for the sensation of hue differences, below which objectively present differences can no longer effect differences in sensation. It is possible, therefore, continually to fill the entire hue circle, i.e., in such a manner that nowhere will a difference between two adjacent hues be experienced. The number of hues required for this is about 300.

Complementary Colors. If one moves away from a given hue in the hue circle, the colors become increasingly dissimilar. But as has been mentioned, this does not continue without limit, because the chromatic sequence returns full circle, so that from a certain point on the colors will again become more similar until one returns to the point of origin. If we select the yellow color 69 on Color Plate II as a point of origin, 71 will be less similar to it than 70, 72 still less similar, and at 73 there can no longer be any thought of similarity. In contrast, 74 already begins to approach 69 in appearance, 75 is still closer to it, and 76 can be said to be quite similar.

The same observations can be made for every other full color.

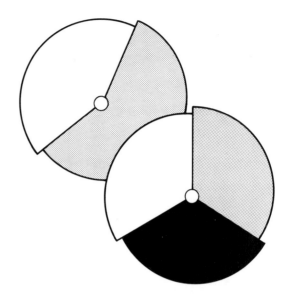

Figure 13. Visual mixtures of colors can be determined by arranging disks as shown here, spinning them on a color wheel and measuring proportions.

Thus, there exists for every hue in the hue circle another that is most different from it. This relationship is mutual. The entire hue circle is filled with such pairs of contrasting colors, which shall be called complementary colors.

The complementary colors are arranged in the hue circle in such a manner that colors of each pair will be located opposite each other. However, the determination of the complementary color of a given hue according to the law of least resemblance is very inexact, as there may be some doubt as to what particular point on the color circle resemblance is least. Yet, exact determinations can be obtained by a mixture process.

Mixture. If one mixes two colors, the hues of which are close to each other, a new color will be created that lies between the two components, i.e., it resembles each of them more than the two resemble each other. This applies to the mechanical mixture of pigments as well as to the optical mixture of colors such as might be spun on a revolving wheel. (See Figure 13.)

Here, a new circumstance is noted, which does not exist with the gray colors. In each mixed color the chromatic quality is less pronounced than in the components; the mixed color is duller or less pure. (See Figure 14.) This change is the more marked, the farther distant the components are from each other in the chromatic circle. Finally, one arrives at a pair of colors that in (visual) mixture will yield pure gray. These are the least similar colors or complementary colors, which can thus be exactly determined according to this process. We therefore define:

Complementary colors are colors which in an optical mixture will yield a neutral gray.

If the complementary color is passed, there are obtained again increasingly purer mixed colors.

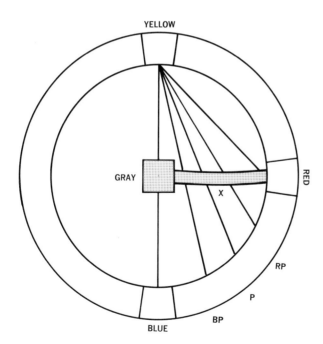

Figure 14. When pure colors are mixed together (visually) the more dissimilar they are the more the resultant combination will approach gray — path X above.

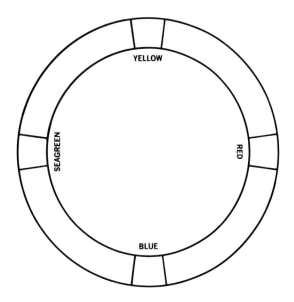

Figure 15. The color circle has yellow, red, blue, seagreen as its fundamental colors.

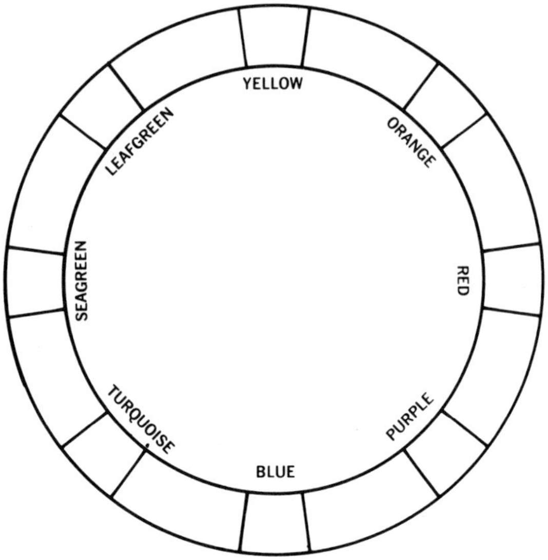

Figure 16. With orange, purple, turquoise, and leafgreen added (to Figure 15) the color circle has 8 principal hues.

To distinguish the most important complementary colors, the chromatic circle on Color Plate II should be studied. (Also see Figure 17.) Here, the colors have been arranged in such a manner that the complementary colors are situated exactly opposite one another. We find the following pairs:

      Yellow — Ultramarine Blue
      Orange — Turquoise Blue
      Red — Seagreen
      Purple — Leafgreen

These eight colors are called the principal colors; yellow, ultramarine blue, red, and seagreen are also called the fundamental colors. (See Figures 15 and 16.) The earlier assumption of 6 principal colors and 3 fundamental colors is incorrect.

Division of The Hue Circle. Connecting each pair of complementary colors on the hue circle are two semicircles. Starting with the pair yellow-blue, the first semicircle (to the right on Color Plate II) will be composed of yellow, orange, red, purple, to ultramarine blue, with all intermediate variations. In the other semicircle (to the left) are ultramarine and turquoise blue, plus seagreen and leafgreen with their variations.

Each half is further arranged according to the principle that any two colors mixed in equal parts will determine the color situated exactly in the center between them. In the circle so created, 24 equidistant steps are chosen which are given numbers from 1 to 24. (See Figure 17.) Thus, there will be three steps for each principal color, namely, 1 to 3 for yellow, 4 to 6 for orange, 7 to 9 for red, 10 to 12 for purple, 13 to 15 for ultramarine blue, 16 to 18 for turquoise blue, 19 to 21 for seagreen, 22 to 24 for leafgreen.

The individual steps are close enough to each other so that for practical purposes no other hues are needed between them. On the other hand, they are far enough apart so that one can easily differentiate between them.

For a complete visualization, a twenty-four part hue circle is reproduced on Color Plate II. The hue number is given for each color.

These 24 colors have been formed by the 8 principal colors of the circle of Figure 16, in that each of these was split into three equidistant steps. If these are called first, second, third yellow, first, second, third orange, etc., one obtains the following table:

	1st	2nd	3rd		1st	2nd	3rd
Yellow	1	2	3	Ultramarine Blue	13	14	15
Orange	4	5	6	Turquoise Blue	16	17	18
Red	7	8	9	Seagreen	19	20	21
Purple	10	11	12	Leafgreen	22	23	24

It is worth remembering these numbers, names, and corresponding hues, as they will be used continually.

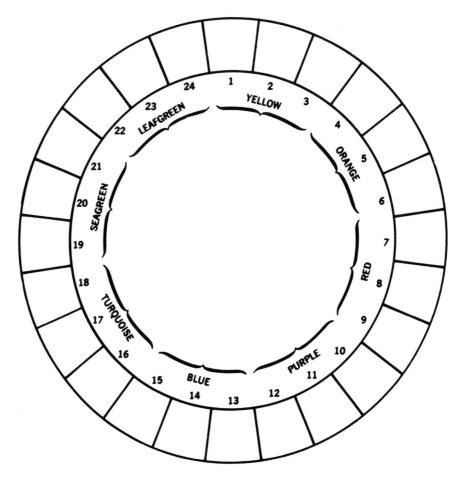

Figure 17. The complete color circle. Also see Color Plate II.

The 8 principle colors are identified as the 2nd colors located in the middle of each group of three; namely 2 for yellow, 5 for orange, 8 for red, 11 for purple, 14 for ultramarine blue, 17 for turquoise blue, 20 for seagreen, 23 for leafgreen. These are the colors plotted on Figure 17.

Standardization of Hues. As the 24 hues are sufficient for practical purposes they can be used as color standards, especially since they have been chosen in such a manner that they are equidistant from one another and they uniformly cover the hue circle.

True, the beginner may think that the colors in turquoise blue and seagreen are closer to each other and more difficult to distinguish than the others. This is due to the fact that this area of the color circle is very little known to us, since these colors hardly occur in nature. Colors of flowers do not as a rule go beyond the third ultramarine blue, and the colors of leaves begin only with leafgreen; thus, turquoise blue and seagreen are missing in them. The same colors are absent in butterflies, and among birds only the rare and shy kingfisher sports them. All the other hues are represented in nature amply and in great variety and are therefore well known to us.

THIRD CHAPTER

Light Clear and Dark Clear Colors

Full Colors. Colors expressing only hue, without the admixture of white, gray, or black, are called full colors: they are ideals which cannot be produced as actual pigments or dyes, for they always contain some amount of white and black in their makeup in addition to the full color. The colors used for Color Plates I and II, and which approach full colors as closely as was technically possible, still contain about 5 percent white and as little black as possible.

The Light Clear Series. As in Figure 18, colors created by the addition of white to the full color, such as examples 41 to to 44 on Color Plate I, are called light clear colors. They have the following characteristics.

Small amounts of white which are added to the full color, e.g., by means of the color wheel (Figure 13), strongly affect appearances. The result is similar to the addition of small amounts of white to black. But this influence depends on the hue. Changes are most marked with blue and purple and are weaker with yellow. This is related to the color's own luminosity, which is strongest for yellow, weakest for blue.

Conversely, small amounts of full color added to white cause but little difference. At first, the white only appears somewhat modified or darker. It is possible to recognize the hue quality clearly only when the quantity of full color added exceeds 10 per cent. Then the hue comes to the fore. Here, too, variations are observed in the hue.

Thus, similar relations apply to mixtures of white and full color as apply to mixtures of white and black.

Besides, gradations in the light clear series run between white and full color just as constantly as between white and black, and there exists here, too, a distinctive threshold which is dependent on the same conditions as for the achromatic grays. Therefore, there are several hundred distinguishable gradations between white and each full color.

Standardization of Light Clear Colors. In order to bring order and a lucid view to the immense variety of chromatic colors, it is necessary here, too, to standardize them, as has been done with the gray series. This work is facilitated in that in both cases very similar rules apply.

To effect gradations in the light clear colors, which are to appear equidistant, the relationship must be determined according to a geometrical progression, just as was the case with the gray series. In this instance (the light clear

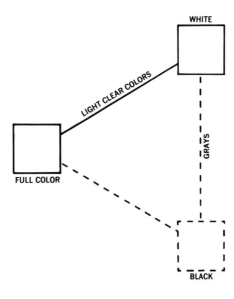

Figure 18. Light clear colors are formed by mixing white with pure hue (full color).

series), the full color will take the place of the black, while the white proportions will be the same as with grays. Accordingly, the steps in the light clear series are marked with the same letters a, c, e, g, i, l, n, p, etc., wherein the letter a means white (without chromatic color), just as in the achromatic series. The palest color, i.e., the one containing the most white, will be designated c, the deepest one containing the least white being designated p. Yet p is by no means the deepest color that can be created, because on wool, silk, and especially velvet it is possible to get deeper. Only the nature of the coloring agents effect a practical limit here.

(Editor's note. It will be helpful at this point to refer to the Ostwald color triangle, Figure 29, found in the next chapter. In all Ostwald designations which have the letters a, c, e, g, i, l, n, p, the *first* letter in all cases signifies the white content in the color, and the *second* letter signifies the black content — as traced for the gray scale of Figure 11.)

Designation of Light Clear Colors. From each full color, a series of light clear colors leads to white. As we have 24 hue standards on our color circle, 24 series of standardized light clear colors are created, which are characterized first by the *number* of the hue and second by a *letter* which indicates the white content. Thus, there belong to the second ultramarine blue 14 (see color circle on Color Plate II) the light clear derivatives which have a white content respectively of c, e, g, i, l, n, p. It is thus apparent that these colors should be designated 14c, 14e, 14g, etc. To make certain that these are light clear, i.e., nearly black-free colors, we add a second letter a which signifies black content. By referring to the gray scale of Figure 11, the letter a signifies a pure white. Therefore where it is used as a *second* letter, black would be absent in the color. Thus, the said colors would be designated as 14ca, 14ea, 14ga, 14ia, 14la, 14na, 14pa.

As each full color thusly obtains 7 light clear derivatives, the number of light clear norms is $7 \times 24 = 168$.

The light clear series red 8ca to 8pa and seagreen 20ca to 20pa are shown on Color Plate III and are numbered from 101 to 107 and 108 to 114.

Production. The painter is acquainted with the light clear series, as they are created when an exceptionally pure pigment is mixed with increasing quantities of a white. In painting with watercolors, similar series are obtained when the transparent pigment is applied to white paper in thin layers. It is found that these series, too, (as in all series of chromatic colors) are constant.

A more detailed examination will show that when mixing a color with white or diluting it with water, the hue of the pigment is often affected by some change. In orange, for example, the hue will shift toward yellow with increased dilution.

However, red and purple change only little; blue becomes greener upon dilution, green around 21 remains steady, yellow-green becomes bluer. These are the general conditions, which are subject to influence and change by the particular nature of the pigments. Also, the same pigment will act somewhat differently when it is mixed with zinc white than when it is applied as a transparent glaze over white; in the former instance, the mixture generally turns out bluer. This is due to the fact that all "dulling agents" (to which zinc white belongs) cause a blue effect when they are spread over a dark ground. In mixtures containing zinc white this is present through the granules in the pigment.

Swatches 115 to 118 on Color Plate III show the colors of a blue pigment having increased dilutions with white, while swatches 119 to 122 show the adjacent hues of the hue circle. (Also see Figure 19.) One recognizes that the deviation in hue becomes greater the further the dilution goes. Between 115 and 118 it amounts to the equivalent of two steps on the color circle toward seagreen.

Change of Appearance. The shifting of hue in a colored pigment when it is diluted or mixed with white — which represents almost all the experiences we have gained with the light clear series — has strongly influenced most of

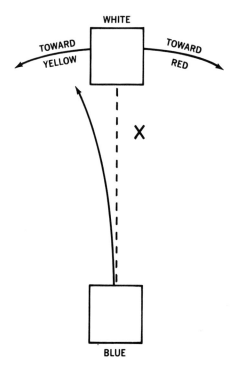

Figure 19. When white is added to some blue pigments, the tints will become greenish in tone. However, a true light clear series will lie along line X.

us regarding visually true color scales. When we observe a visually accurate light clear series, we tend at first to judge the dilutions of orange, purple, and ultramarine as being too red, and those of turquoise blue and yellow-green as being too blue. But we must first learn which are the correct series.

Most noticeable is the deviation around point 13 of the chromatic circle (ultramarine blue). At first, one does not wish to believe that the series 123 to 127 on Color Plate III does indeed represent the same hue and that it only contains increasing amounts of white. We tend to regard the lighter colors as being quite red. In actuality, however, all these steps (123 to 127) require the same yellow 1 to achieve pure gray by optical mixture, and the lighter colors are obtained from the darker ones when they are optically mixed with white.

The Dark Clear Series. As in Figure 20, if one mixes full colors with increasing quantities of black, dark clear series are created, and these follow similar rules as to the light clear series. However, while light clear variations of pure hues can be accomplished technically with good approximation, this is not the case with the dark clear ones. This is due to the fact that there are no black paints which do not reflect some visible quantities of light, as we have discussed in connection with the black-lined box described in the first chapter. Similarly, all other dark-colored paints will reflect several percentages of white light, and this will dull the purity of dark clear colors in a very noticeable way.

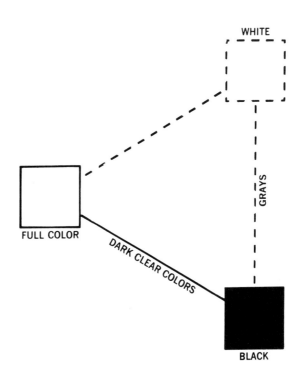

Figure 20. Dark clear colors are formed by mixing black with pure hue (full color).

Dark clear colors can best be seen in stained glass windows, especially old church windows. Here white is not contained, or if it is, the extent is very slight. We find that where the colors of the stained glass are darkened step-by-step to obscurity (intentionally through burning of a black powder or unintentionally through accumulations of dust and soot) dark clear colors will appear which lead from the pure color of the stained glass to full black. In these instances there is little or no admixture of white. In paints and fabrics, however, white is usually present because of the uncolored light reflected by the surface.

As related to paints (or printing inks), the dark clear colors can be realized to an incomplete degree only; swatches 128 to 132 on Color Plate III, a red hue, represent the best approximation that could be achieved with available means, yet all steps still contain about 04 to 02 percent white.

Graduation of Dark Clear Colors. A similar rule as governs the graduation of the achromatic and the light clear series also applies to the dark clear colors. Only that here the full color will take the place of the white and thus be mixed with black. To create an equidistant series of dark clear colors of the same hue, the portion of the full color must be decreased according to a geometrical progression. From this it follows that it is necessary to add large quantities of black to the full color before much of a change becomes apparent, while even small quantities of full color when added to black are clearly discernable.

In relation to white (in light clear colors) the full color behaves like black, while in relation to black (in the dark clear colors) the full color behaves like white.

Change of Appearance. While in the light clear series the blue area showed the strongest changes in appearance, with the dark clear colors the strongest changes are in yellow. We usually do not even describe darkened yellow as yellow, but as olive-green. (See Figure 21.) Swatches 133 to 138 on Color

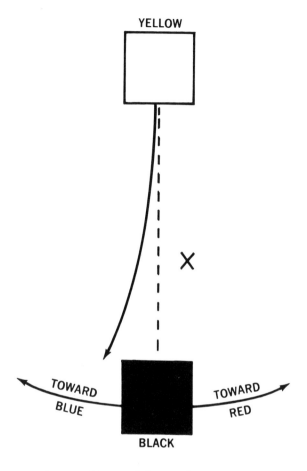

Figure 21. When black is added to any yellow pigment, the shades will become greenish or olive. A true dark clear series will lie along line X.

Plate III are approximately dark clear yellows of hue 2 having increasing amounts of black and 04 percent white. The steps also give the appearance of a considerable change in hue towards the green side with the increased contents of black, just as blue swatches 123 to 127 appeared to change to the red side with increasing contents of white. However, they both are complements, blue to yellow, and yellow to blue, and will neutralize each other if mixed visually.

Here, too, the reason for the error in judgment is that in day-to-day life we have little opportunity to get acquainted with visually true dark clear series. But that there exists an inner connection, recognized by all of us, comes to the fore when such series are used in a decorative way; their harmonic charm is then apparent.

FOURTH CHAPTER

The Muted Colors

The Variety. If a color contains, in addition to a full color, simultaneously white and black, it is called muted and becomes part of a shadow series. If one attempts to imagine all the muted derivatives that could originate from a given full color, one realizes that one series is not sufficient for this, a form in which we have thus far been able to arrange other variations of colors toward white or toward black. It is, of course, possible to create series of muted colors between a full color and every step or value of gray (for example, gray i on Figure 11). However, since every gray will make further series with the same full color and all these series together belong to the same full color, another and more logical form of presentation must be chosen. (See Figure 22.)

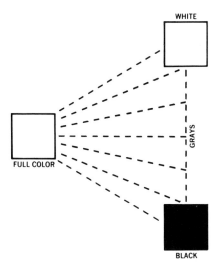

Figure 22. Muted colors could be arranged as indicated here. Ostwald preferred the arrangement of Figure 29.

The Monochromatic Triangle. Such a logical presentation can be arrived at in the following manner. The perpendicularly placed gray series WB (W = white, B = black) is placed opposite the full color C, and the point C is connected with the two extremities of the straight line WB, as on Figure 23. Thus, the triangle CWB is created. In it, a line can be drawn from the full color point C to every gray, i.e., to every point of side WB, on which line all graduations between the full color and the appropriate gray can be accommodated. (See Figure 22.) Thereby all possible mixtures of the full color with any gray, at any ratio, can be given a location inside the triangle. The triangle therefore encompasses all possible such mixtures, i.e., all possible muted colors deriving from the selected full color, white, and black.

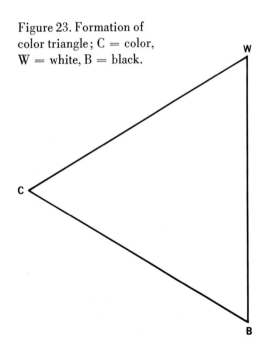

Figure 23. Formation of color triangle; C = color, W = white, B = black.

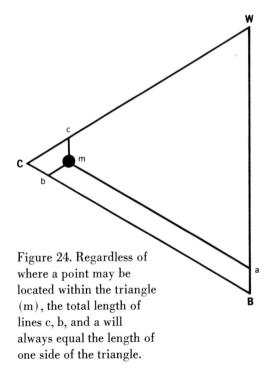

Figure 24. Regardless of where a point may be located within the triangle (m), the total length of lines c, b, and a will always equal the length of one side of the triangle.

Because all these derivatives of the same full color are present (i.e., they all have the same hue), such a triangle is called a monochromatic triangle.

The Color Equation. In general, all colors consist of a part of C full color, a part of W white and a part of B black. If these parts are expressed in percentages, we get for each color the equation:

$$C + W + B = 100$$

In muted colors, all three elements, C, W, and B have finite values. In light clear colors which contain no black, $B = 0$; in dark clear colors which contain no white, $W = 0$. Finally, in achromatic colors $C = 0$.

If for a given color the hue and two of the three values C, W, and B are known, the visual composition of the color is complete. For the third quantity is then also known, as it can be calculated with the aid of the equation $C + W + B = 100$. It is merely appropriate, for the characterization or identification of colors, to state the parts of white and the parts of black that are present in a particular color (as mixed with the full color). Accordingly, if one then wants to know the parts or quantity of full color, it can be calculated on the basis of these data, with the aid of the formula $C = 100 - W - B$.

(Editor's note. Figure 24 — which is in Ostwald's book, but which is not discussed in the text — clearly explains the principle of Ostwald's triangle. When any step or point is located within the triangle, if one line is drawn vertically, one toward the left and one toward the right, the total length of these three lines will always equal the length of any side of the triangle! In other words, $C + W + B = 1$. Any color, regardless of its variation, will be found to be visually composed of a full color, white, and black. The portion or quantity of any one element cannot be increased without at the same time decreasing portions of the other elements.)

Particular Series. In the monochromatic triangle we know what lies along all three sides. On side WB is the achromatic or gray series; side CW has the light clear, and side CB has the dark clear series.

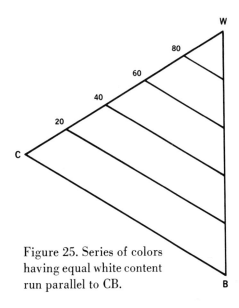

Figure 25. Series of colors having equal white content run parallel to CB.

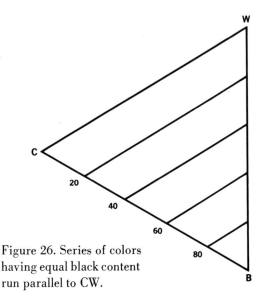

Figure 26. Series of colors having equal black content run parallel to CW.

Parallel to these sides, there run in the triangle certain other series, the colors of which stand in particularly close relationship to each other, as in them one of the quantities W or B remains unchanged in the color equation. We thus obtain the equal whites and the equal blacks.

The equal white series run parallel to the lower side CB, in which the dark clear series are situated (Figure 25). At the lower extreme, CB, white content is zero throughout. Here is the extreme equal white series. As the series or scales approach W, they become shorter. More white is added to each row in uniform amount until pure white (W) is reached.

The equal blacks (Figure 26) run parallel to the upper side CW of the triangle. At the upper extreme, CW, are situated the light clear series. Here black content is zero. As the scales approach B, they also become shorter, and each row has a uniform and increasing proportion of black until pure black (B) is attained.

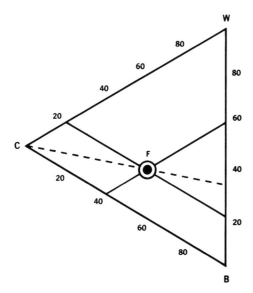

Figure 27. Variations of a given hue can be accurately plotted on the color triangle.

Fixing the Position of the Colors. If the white and the black contents of a color are known, its position on the monochromatic triangle can be determined by plotting the white according to the white content on the gray scale and the black according to the black content on the gray scale. The intersection of the two lines will be the position of the color. Thus, e.g., (Figure 27) a color of blue hue 14 with 20 white and 40 black occupies the position F in the monochromatic triangle, blue 14, F being the intersection of white quantity 20 and black quantity 40. (Editor's note. This may be somewhat confusing. The 40 black content mentioned by Ostwald is determined by subtracting 60 from 100 on Figure 27, inasmuch as on this Figure white quantities alone are indicated. Also, by analysis, a blue with 20 white and 40 black would theoretically contain 40 full color, all these numbers equalling 100. $C + W + B = 100$.)

Standardization of the Monochromatic Triangle. With well-ordered series of light clear colors and dark clear colors, the areas of monochromatic triangles are now filled with color variations that pass continuously into each other. The distinguishable number could run into the hundreds of thousands. Here, too, we must introduce standards to create order and clarity.

Colors in general are obtained by starting out with the already standardized achromatic side of the triangle WB, which is divided according to the steps of the gray scale illustrated in Figure 11. By drawing parallel lines from the graduated points to the two other sides, CW and CB, the triangle is divided into 28 rhombuses, covering all the chromatic derivatives of a given full color. In addition, 8 half rhombuses are formed along the side WB, and these represent the achromatic steps (Figure 28).

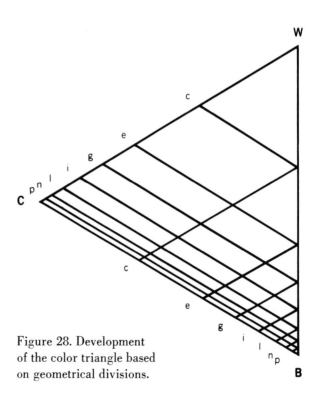

Figure 28. Development of the color triangle based on geometrical divisions.

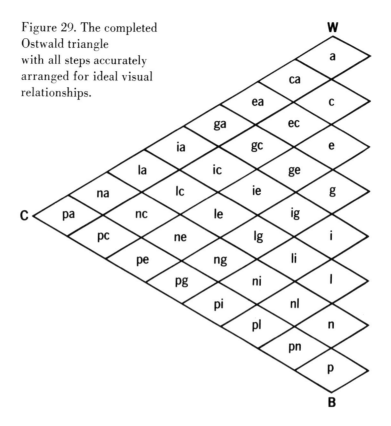

Figure 29. The completed Ostwald triangle with all steps accurately arranged for ideal visual relationships.

Because the intervals toward black get smaller, due to the geometrical progression of the gray steps, the rhombuses are unequal in size. If equal steps *of sensation* are to be allotted equal-sized spaces, the triangle must be stretched until the intervals on side WB become the same. Then, all rhombuses will also be of equal size and the triangle will appear as in Figure 29. Here as well the rhombuses along side WB have been made complete. Such a triangle yields the desired norms or steps, and there is unity and neat visual sequence in all directions.

Details of Standardization. In our new triangle, Figure 29, the equal white series run parallel to the lower side of the triangle and the equal black series run parallel to the upper side. In all instances, the white content (first letter) and black content (second letter) are adjusted so that they will correspond to the letters a, c, e, g, i, l, n, p of the achromatic series (Figure 11). In the equal white series (parallel to CB) the bottom row is called the clear darks. The colors in this row, having p as the first letter, contain no white — because p on the gray scale is a pure black. The steps in the next equal white row, na to n, contain a uniform n amount of white as traced from the gray scale, and so forth.

Correspondingly, the equal blacks run parallel to the upper side of the triangle. The uppermost row or scale (pa to a) has the light clear colors. Here the black content is approximately zero, because a on the gray scale is a pure white. Then follow other rows having equal black content as on the gray scale — rows leading to c, e, g, i, l, n. Color pa is a full color because it contains neither white nor black.

The above points are graphically shown in the triangle of Figure 29. Every rhombus contains two letters, of which the first indicates the white content and the second the black content of the color found in the rhombus (as traced from the gray scale). Each rhombus has a different pair of letters, thus representing a different color variation. There could also be a rhombus for every conceivable color. In this manner, the triangle could contain all possible variations or derivatives of a selected full color up to white content p.

The single letters a, c, e, g, i, l, n, p on the right side of the triangle appropriately indicate the white and black content of the achromatic or gray colors. As these are already sufficiently identified (see Figure 11), it suffices to enter the simple letters, as has also been done previously.

On Color Plate IV there is shown the monochromatic triangle of yellow hue 2. For technical reasons one step each had to be skipped, so that only the steps a, e, i, n have been executed.

Diverse Hues. A triangle of the kind just described belongs to every hue or color. As standardization in this book has yielded 24 hues, there are thus 24 monochromatic triangles.

In each triangle are contained 28 chromatic variations separated from black and white. The number of chromatic hue steps is therefore 24 × 28 = 672. To these must be added the 8 achromatic ones, bringing the total number of color variations to 680.

This holds true subject to the provision that we stop the series at p. If technology will permit the manufacture of richer colors, pigments, or dyes with less white or black in them, the number of variations will also grow, and at a steep rate at that.

The Color Symbols or Notations. The pairs of letters which indicate the white and black content of colors are also the best means to name them briefly and noninterchangeably. In addition to this, one needs only to add the hue number, which is the same for each monochromatic triangle. Thus, 8 le represents a color of hue 8 (second red), white content 1, and black content e. (This would be a tone of rose.) We call such combinations, composed of the hue number and the letters for white and black content, color symbols or notations. They are to color organization what notes are to music. They serve not only to name various colors briefly and precisely, but they also provide a tool for arranging them and finding from among them those that agree with one another, the harmonious colors.

Color symbols for all variations of the *same hue* have the same number. In all colors with the *same white content* the first letter of the color symbol is the same, in all colors with the *same black content*, the second letter is the same.

FIFTH CHAPTER

The Color Solid

The Totality of All Colors. The hue circle comprises all hues; each monochromatic triangle comprises all derivatives of a given hue as visually mixed with black and white. In addition, there is no color variation that cannot be plotted and identified on the triangle. Therefore, if all the triangles are put together around a hue circle, one obtains a display of the entire color world.

The Color Solid. Such an arrangement is created when all monochromatic triangles are arranged in proper hue sequence around a common axis in such a manner that the triangles extend outward in all directions, as in Figure 30. (Also see Figure 10.) They will then form a double cone (Figure 31) whose peaks will have white at the top and black at the bottom, while the full colors are around the circumference, forming the hue circle. The upper surface of the cone contains the light clear colors, the bottom one the dark clear colors. The muted (grayed) colors are within the form and become paler the closer they come to the upper apex, more blackish the closer they get to the bottom, and purer the closer they approach the outer circumference.

This double cone, which comprises the entire world of colors, is called the color solid.

The Steps in the Color Solid. In place of a continuous color solid in which color variations are imperceptible, there is practical need for a standardized color solid, which consists of a finite number of colors, arranged according to equal intervals of sensation. The discussion of the monochromatic triangle (Figure 29) directly yields the order of the color steps in the color solid.

It must be remarked that the lower outer cone surface, which contains colors which lack a white content, does not include the absolutely lowest shades that

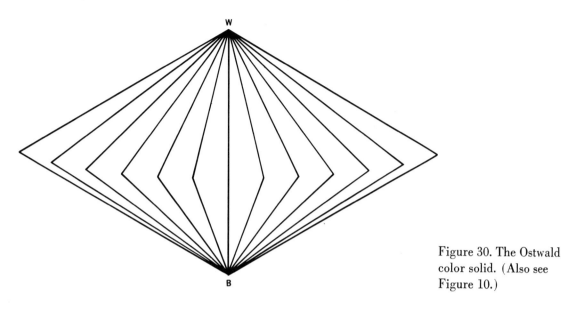

Figure 30. The Ostwald color solid. (Also see Figure 10.)

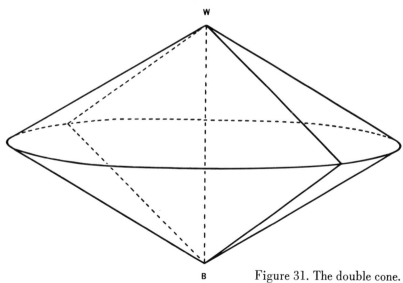

Figure 31. The double cone.

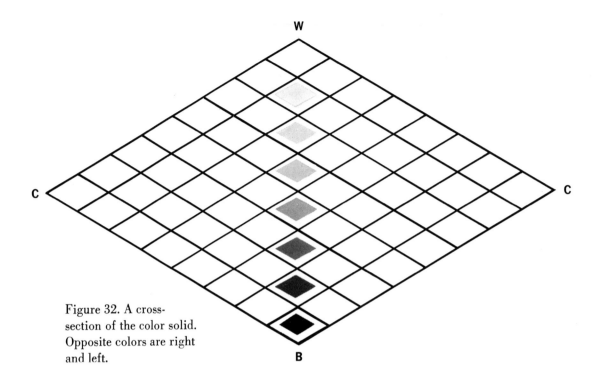

Figure 32. A cross-section of the color solid. Opposite colors are right and left.

might be produced. Rather, even deeper colors could find their place here just as soon as richer colors are developed as pigments, dyes, or other coloring materials.

Principal Sections. To gain a more exact view of how the color solid actually looks, it is appropriate to think that it was sliced in certain directions, so as to uncover its interior. Most instructive are such sections which divide the solid into two halves, straight down from white apex to black base, as in Figure 32. Thereby, two monochromatic triangles are laid bare in each instance, and their hues are at opposite points on the chromatic circle. They are thus complementary colors.

The illustrations of cross sections of the color solid on Color Plates V and VI show four such "principal sections." Color Plate V contains the principal colors yellow 2 and blue 14 at the top of the page, and orange 5 and turquoise 17 at the bottom. Color Plate VI contains the principal colors red 8 and seagreen 20 at the top of the page, and purple 11 and leafgreen 23 at the bottom. Here, too, every second letter has been omitted in the series a, c, e, g, i, l, n, p. Thus, the basic series a, e, i, n has been presented, giving to the chromatic color variations the symbols ea, ia, ie, na, ne, ni. (See Figure 33.) Instead of the 64 rhombuses of a complete principal cross section, only 16 appear here in each instance, and the "small color solid" of Color Plates V and VI is formed by 48 chromatic and 4 achromatic colors.

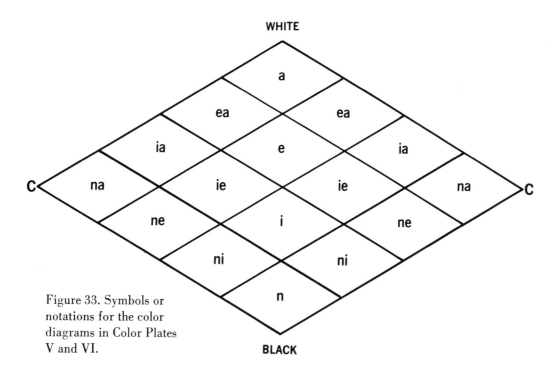

Figure 33. Symbols or notations for the color diagrams in Color Plates V and VI.

In each principal section the achromatic axis can be recognized, running in the center from the top down. From it and towards both sides, the equal whites ascend obliquely upwards and the equal blacks drop downward obliquely. The closer the colors are to the axis, the less pure they are.

The Shadow Series. In addition to the equal whites and equal blacks, which run parallel to the sides of the triangle, there is another important class of correlated colors. They run from top to bottom parallel to the achromatic main axis, WB, and diminish the more they approach the outside corners of the rhombuses. (See Figure 34.)

These series are called shadow series because they contain colors that have the effect of being created by means of shadowing or reduced illumination.

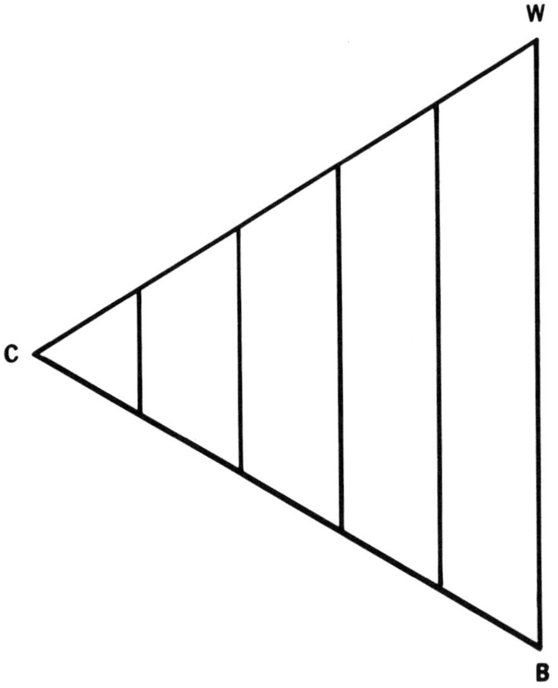

Figure 34. The shadow series. Scales of colors that run vertically parallel to WB appear to have equal hue content.

They are of particular importance because we can derive from them how the form of a given color is to be shaded.

The shadow series therefore contain colors of apparent equal purity or hue, and are thus also called equal purity (series). The achromatic (gray) row in the axis of the color solid has zero hue purity.

(Editor's note. Ostwald placed great emphasis on the beauty of the shadow series. [See Figure 29.] There are 6 vertical scales, ca to pn having 7 steps, ea to pl having 6 steps, ga to pi having 5 steps, ia to pg with 4 steps, la to pe with 3 steps, and na, pc, with 2 steps. He remarked that the magnificence of the *chiaroscuro* style in painting perfected by such masters as Leonardo da Vinci and Rembrandt was largely based on the subtle gradations of the shadow series.)

Equal Value Circles. Just as the purest colors of our color solid, neatly positioned around the circumference of the double cone, form a hue circle such as on Figure 17 and Color Plate II, this takes place also for every other area or step of the monochromatic triangle. The circles so located will then be larger or smaller, depending on their position within the solid. But they are of equal size for all members of a shadow series, being equidistant from the main axis in all cases.

In each of these circles *the white and black content is equal everywhere;* the corresponding areas or steps of the triangle carry the same letters. The colors of such circles are called "wertgleich" (equal value) and each of these can be identified by the corresponding pair of letters. Thus, there are circles ca, ea, ec, ga, etc., down to pn, a total of 28 "equal value" circles. These can be located on Figure 29. As each contains 24 colors, there is found the total number of $24 \times 28 = 672$, the same as for the entire color solid itself.

On Color Plate VII are illustrated two "equal value" circles, ea and ni (abbreviated to the 8 principal colors), of which one contains much white, the other much black.

It is possible, therefore, also to organize the entire color solid in terms of 28 "equal value" circles instead of in terms of 24 monochromatic triangles. Each of these two arrangements expresses certain important relationships; in one instance an arrangement by hue, and in the other an arrangement by equal white and black content. Both arrangements are present in the color solid simultaneously, because all hue circles have their centers in the axis of the color solid and all triangles are located in its principal sections. This is of particular importance to the problem of color harmony.

SIXTH CHAPTER

The Harmony of Colors

The Principle. Experience has shown that certain combinations of different colors have a pleasant, others an unpleasant or indifferent effect. The question arises, on what does this depend?

The answer is: such colors will have a pleasing effect, between which there exists a lawful relationship, i.e., order. If this is absent, they will have an unpleasant or indifferent effect.

We call color groups that have a pleasing effect harmonious; we can therefore establish this basic law:

$$\text{Harmony} = \text{Order}$$

To find all possible harmonies, one must study all the possible orders (arrangements) of the color solid. The simpler the order, the more illuminating or convincing is the harmony. Of such orders we have found two primary ones, namely, the equal value (equal white and equal black) hue circles and the monochromatic triangles. The circles result in equal value harmonies of *different* hues, while the triangles yield monochromatic harmonies of *different* white and black content of the same hue.

Gray Harmonies. The simplest application of the principle is present in the standardized achromatic colors, as they are arranged by equal intervals (Figure 11). If a form (a sample, clothing, artwork) is to be decorated in different steps of gray, they should be selected from the standard series so that they will have equal intervals. For example, a, c, e or g, i, l (simple intervals), and also a, e, i or g, l, p (double intervals), and a, g, n or c, i, p (triple intervals). (See Figure 35.)

To such gray harmonies will fit chromatic colors whose color symbol contains the same letters. Thus, to a gray harmony in g, l, p one may take chromatic colors, the color symbols of which end in lg, pg or pl (see Figure 29); the hue may be as desired. The gray-chromatic harmonies thus created are exceptionally beautiful.

Figure 35. Neatly spaced intervals anywhere on the color solid lead to harmony.

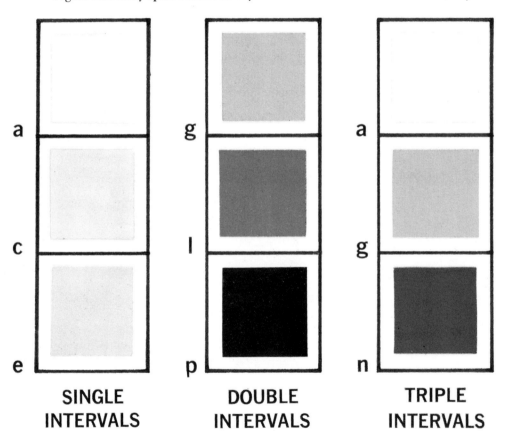

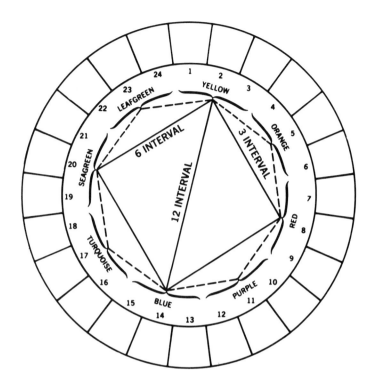

Figure 36. Neatly spaced intervals will lead to harmony either among pure hues or among equal-white or equal-black colors found on equal value circles.

Equal White and Equal Black Harmonies. Harmonies of different hues are found in equal white and equal black circles and are therefore also called equal white and equal black harmonies (or equal value harmonies). The most intense and most striking harmonies can be found when one advances or retreats from the selected hue by 3, 4, 6, 8, or 12 steps (Figure 36). It is also possible to take such steps repeatedly in succession and then return to the point of origin with 8, 6, 4, 3, 2 steps. Only interval 9 does not give a simple division of the circle.

Well known are the complementary color pairs with an interval of 12, whose effect becomes more intense the purer the colors. On Color Plate VIII the complementary color pairs yellow 2 and blue 14, and red 8 and seagreen 20 have been combined out of the circles na, ne, ni. (See Figure 29.) One can see how the effect becomes softer with decreasing purity. (Swatches 229, 230, 231, 232, 233, 234.)

Besides complementary color pairs, triads with interval 8 are often used, which are obtained by ordinary division by three of the hue circle. As triad colors in the past have mostly been wrongly defined, the four true and most important triads (from the equal value circle ne) are correctly shown on Color Plate VIII.

All colors in black circles at the bottom of Color Plate VIII are at step ne on the color triangle (Figure 29). All are triad combinations of principal colors spaced 8 intervals apart on the color circle (Figure 17). Swatch 235 combines yellow 2, purple 10, turquoise 18. Swatch 236 combines orange 4, purple 12, seagreen 20. Swatch 237 combines orange 6, blue 14, leafgreen 22. Swatch 238 combines red 8, turquoise 16, leafgreen 24.

To such equal white and equal black chromatic harmonies only those achromatic colors will be suitable which correspond to the letter of equal white and equal black circle from which the harmony was taken. For example, to the harmony 2, 10, 18 ne (swatch 235 on Color Plate VIII) the gray colors e and n are suitable.

Monochromatic Harmonies. The general and balanced order of the monochromatic triangle does not allow for haphazard color arrangements. Rather, one is wise to restrict oneself to individual series therein, such as the light clear series (equal white) or the dark clear series (equal black).

However, there is particular beauty in the shadow series. (See Figure 34.) Colors in these vertical rows yield the so-called shade-within-shade harmonies. In these series full colors are mixed with white or black or both (gray). However, since, in this connection, hue shifts, previously mentioned, often occur, designs or works of art created with *true* shadow series colors will have a very harmonious and "artistic" effect. Correct (real) shadow series colors can be found in vertical rows a to n or e to i on any of the triangle cross sections of Color Plates V and VI.

In addition, it is possible to arrange equal white and equal black harmonies, using neatly spaced intervals within the triangle as well as moving from one hue to another at the same time, also in neat intervals.

Combined Harmonies. By using the various special laws of harmony simultaneously, many diverse relationships can be obtained. These often are imbued with particular charm. In this manner, it is possible to create more complicated orders of beauty step-by-step, and the already very large realm of color harmony could be expanded to infinity.

Thus, it is possible, in an equal white, equal black two-hue (harmony), to replace one of the colors (or both) entirely (or in part) by one (or several) members of the corresponding shadow series. For example, if color harmonies were arranged along row la to li on Figure 29, or along row pe to ge, colors along the shadow series ga down to pi could be introduced. This rather obvious idea alone offers a manageable wealth of possibilities with a single pair of hues. But these can be repeated with all equal valued harmonic pairs present within the color solid.

Conclusion. Of the many possible laws of color harmony only the simplest ones could be mentioned here. Their application expands the already vast realm of color harmony into the unimaginable. It would require the combined work of many hands and many years just to create the most important among them in their simplest form and to exhibit them. A beginning has been made with the *Harmothek*, which has been published since 1926. It is a collection of executed harmonies with explanatory text. (Editor's note. No English version of the *Harmothek* was ever produced.)

If one were to ask which colors would combine best with a given hue, the following reply would result. First in harmony would be other colors which have the same value or white and black content, i.e., to be precise, such colors as have the same letters on the color triangle (Figure 29) but which preferably are spaced at intervals of 3, 4, 6, 8, or 12 on the color circle (ring A on Figure 37). Second in harmony would be colors which have the same white content (row B on Figure 37) or the same black content (row C on Figure 37). Or colors which have the same apparent hue content (the shadow series, row D on Figure 37). Of a total of 38 steps on the color triangle one would use in each particular case only a few, according to the particular law of harmony that one wishes to apply.

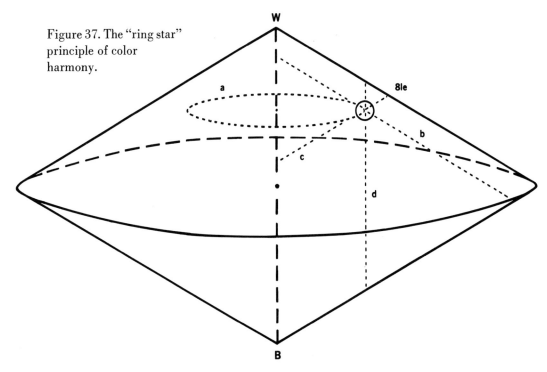

Figure 37. The "ring star" principle of color harmony.

All the harmonious colors just described are located in the color solid in a figure composed of an equal white and equal black circle (A on Figure 37), colors of equal white content (B), equal black content (C) and equal hue content (D). All this forms the ring star shown in perspective in Figure 37. Thus, for example, for a color such as red 8 le there will be first other colors in the circle le, especially purple 11 and 12, blue 14, turquoise 16 and 17, seagreen 20 le, and orange 5 and 4, yellow 2, leafgreen 24 and 23, seagreen 20 le; there would be red 8 ga, ic, ng, and pi in equal purity row; 8 la, lc, lg, and li in the equal white row; and 8 pe, ne, ie, ge in the equal black row, plus grays e and l. (See Figure 29 to locate these steps.)

An Evaluation

In the academic field the work of Wilhelm Ostwald has had more recognition in England and Europe than in the United States. However, in commercial fields, and more recently in the fine art of painting in America, Ostwald has gained increasing acceptance and prestige.

There is little doubt that the sense of order inherent in Ostwald's system (the concept of equal white, equal black, equal hue, and equal value colors) has organized the world of color in a magnificent way. His solid is like a great edifice filled with straight passageways leading in many directions to many chambers. No matter what way one may travel, there are pleasing delights to behold. And the traveler need never fear that he will be lost. He has but to note the spot on which he stands — what hue and what location on Figure 29 — and he can proceed with dispatch to whatever point of the color world he wishes to visit.

As to theories of color harmony, needful in education and training, the student or professional can profit greatly by experimenting with those natural orders of beauty found on the color triangle (Figure 29) or within the color solid (Figures 10 and 30). Choice of *hue* can be more or less optional.

There is beauty to be found in Ostwald's equal value colors, ones that have the same black, white, and hue content. One has merely to form color circles in which all hues have the same notation in Figure 29 and arrange them as desired, as adjacents, complements, triads, etc. (Color Plate VII gives two abbreviated examples of equal value circles.)

There is beauty in colors that follow series having equal black content. These run parallel to CW. The top scale, full color to white, has none but fresh, clean pastel tints.

There is beauty in colors that follow series having equal white content.

These run parallel to CB. The bottom scale, full color to black, has deep rich shades.

There is particular beauty in the shadow series, in the rows of colors that run parallel to WB.

Beyond these principles, there are the ring stars described in the sixth chapter of *The Color Primer*.

Indeed, almost any well related, neat interval combinations plotted *anywhere within the color solid* are more or less automatically assured of harmony and concord. Ostwald saw to this when he perfected his master plan.

In the commercial arts and decorative arts, this writer (Faber Birren) several years ago wrote an article for the July, 1944 issue of the *Journal of the Optical Society of America*, "Application of the Ostwald System to the Design of Consumer Goods." Today many commercial products, textiles, wallpapers, carpeting, are organized in color in accordance with Ostwald's views.

In working with color and people in the styling of consumer goods, it will be noted that most of us see the world of color as Ostwald organized it. Such everyday terms as "pastel tints," "pastel shades," "fall colors," "spring colors," "peasant colors" — most of which have meaning to the average American — are obviously identified with consistent forms of pure colors, or white-containing colors, or black-containing colors, or gray-containing colors, precisely as Ostwald's system has been set up. "Pastel tints" and "spring colors," for example, are to be identified with Ostwald's light clear colors which run from pure hue to white. "Pastel shades" and "fall colors" will be found to resemble his dark clear colors which run from pure hue to black. "Peasant colors" will be the pure hues or full colors of his color circle.

Ostwald was especially conscious of the beauty of his muted colors as found in his shadow series (color scales that run parallel to WB on his triangle). He wrote: "If one examines the earliest examples of shading, as for example in the miniatures of the Middle Ages, the method by which it has been attempted to express the shadow series may be easily discerned. The unmixed pigment is adopted for the deepest shadow, a little white is mixed with it for the middle tints, and much white for the lights. In old painting books up to the time of Cennini we find this method of procedure explicitly described."

As to more commercial products of art, he said, "The same method has served to produce the shading of a color on the yarns employed in carpet manufacture, embroidery, etc., right up to the present day. The shades are dyed with concentrated pigment, and the middle shades and lights with progressively weaker material.

"The low condition of our feeling for color harmony is characterized hardly anywhere or anyhow more clearly than by the fact that this radically false method is being practiced up to this day without misgiving. This error gives rise to the intolerable color effect which even the most talented embroiderers are unable to avoid if they use the shades sold by the shopkeepers, and there are no others to be had."

In this criticism Ostwald dealt with perhaps the most conventional and commonplace means of employing color — to work with a basic series of strong rich dyes or pigments and to handle intermediate tones as simple dilutions or mixtures with white. The stores and the factories of the land are packed with such examples — and pretty bad are the most of them.

Far better is it to think in terms of Ostwald's C-W-B concept and to realize that just as color sequences properly follow the order of the color triangle, so do the emotional fancies of most persons.

In effect, and as Ostwald has insisted, color harmony in commercial products becomes less a matter of convenience in dealing with dyes and pigments than of adjusting colors to visual and psychological phenomena. For colors in nature, in highlight and shadow, reveal themselves as Ostwald states and as his shadow series set forth — not as dyes and pigments happen to behave.

In the fine arts, since the demise of Abstract Expressionism (during which all discipline, knowledge, and training in color were abandoned for an impulsive release of "inner necessities") there has been a new respect for the mysteries of human vision and perception. Artists have taken instruction from the psychological sciences, and the value of a well-ordered and coordinated color system — such as that of Wilhelm Ostwald — becomes of great importance as a source of reference for effective color arrangement.

Three Amerian painters (no doubt among many others) have acknowledged a debt to Ostwald and have profited from his doctrines. The first of these is Hilaire Hiler (1898-1966). Hiler wrote a number of excellent books on color and a minor classic for artists, *The Technique of Painting* (Oxford University Press) which has gone into several editions. A "method" painter of the first rank, Hiler, after mastering Ostwald's concepts of equal value scales, developed even more elaborate scales of his own, at times composing hundreds of subtle variations of a single hue into a dramatic canvas. As Hiler only too clearly agreed — harmony = order.

Hannes Beckmann (see Figure 38) originally studied at the Bauhaus where he became acquainted with Ostwald's system. However, as an American painter and professor of art, he has more recently applied Ostwald's principles in original ways. Indeed, the orderly and beautiful sequences always implicit in his art can to some extent be given Ostwald notations, as described under Figure 38.

Ben Cunningham (see Figure 39) has turned Ostwald's technical inventions in color order to exceptionally dramatic ends. Introduced to Ostwald by Hilaire Hiler, he has used the Ostwald system as a remarkable tool. Cunningham states, "The Ostwald System reflects the human nervous system structure in terms of perception; hence its particular value to me." His object has been to work creatively and imaginatively with a well-balanced color solid, just as a composer can find original expression with a well-tuned musical instrument.

Cunningham has added rare talent, feeling, and skill to create astonishing

Figure 38. Hannes Beckmann, "Shadows Over Yellow." In the original, a yellow of extreme luminosity — and outside the limits of the color solid — is assumed to cast shadows as illustrated. The Ostwald notations used by Beckmann are ga, gc, ge; ia, ic, ie, ig; la, lc, le, lg, li, scaling from yellow 1 toward violet.

illusions of space (as in Figure 39). He has manipulated pigments as if they were light, caused human perception to "see" exceptional brilliance of hue, contrived to have opaque surfaces appear as if transparent. He has inspired the viewers of his art to "see what isn't there," to witness imagined conceptions of invisible infrared and ultraviolet, and otherwise to build a remarkable and abstract world of fancy on the foundations of sound color order.

Figure 39. Ben Cunningham, "The Maze According to the Law." Color organization is essentially three-dimensional. Yet in most forms of painting, sense of space may lack conviction because of the obvious flat plane of a canvas. Note in this unique illustration how the artist has succeeded in inviting the eye to travel into distance.

Any student (or professional) in the fields of art and design can, if he wishes, rely on his own inner feelings and compulsions and work with color as the spirit moves. However, while what he creates may be audacious and even different, it is doubtful if it will represent an advanced step in color expression. Indeed, it may be remindful of a child pounding on the keys of a piano — quite exciting to the child, but dissonant to anyone else.

Except in the case of true genius, which is rare, grand accomplishments in the arts do not follow offhand impulses. Periods of discipline and training are the rule. Japanese watercolors and calligraphic designs often take but a few minutes to complete, yet the artist may have spent years of preparation.

In modern times no one can doubt the outstanding talents of a man like Vincent van Gogh. He was a fast and furious painter. He also was a profound student of color theory. He made this observation. "It is not only by yielding to one's impulses that one achieves greatness, but also by patiently filing away the steel wall that separates what one feels from what one is capable of doing." This takes knowledge and understanding.

References

There is a fairly generous bibliography of English works by Ostwald and about Ostwald, and a few truly distinguished publications.

In England itself, where Ostwald was first recognized and first published (ahead of America), J. Scott Taylor offered two of Ostwald's most important writings: *Colour Science*, Part I, *Colour Theory and Color Standardization*, 1931; Part II, *Colour Measurement and Colour Harmony*, 1933. Both were published by Winsor & Newton. At the same time *The Ostwald Colour Album* was produced under the same auspices. This contained six charts, $8\frac{1}{2}$ x 12 inches, having 680 mounted color chips. In all this, Ostwald's doctrines were completely and impressively revealed.

Also in England, in 1934, O. J. Tonks wrote *Colour Practice in Schools*. This was published in two parts, featured Ostwald's theories, and presented "A Graded Course in Colour Seeing and Using for Children between the ages of five and fifteen." Winsor & Newton was the publisher.

In 1935 the Dryad Press of Leicester produced *A Handbook of Colour* by J. A. V. Judson. This included the Ostwald principles and was "A Text-book for Teachers, Students of Art, and all interested in Colour."

Arthur B. Allen wrote *Colour Harmony for Beginners*, for Frederick Warne & Co., 1936, and *The Teaching of Colours in Schools*, for Winsor & Newton, 1937.

Through these publications, Wilhelm Ostwald became an important influence in color training throughout Great Britain.

In the United States, Ostwald had two champions in Herman Zeishold and Egbert Jacobson. (Faber Birren also spoke highly of Ostwald in *Color Dimensions* published by The Crimson Press in 1934.)

Zeishold, who knew Ostwald's family, helped to further the German's work in America in two studies: "The Development of Color Theory Since Newton," a technical study for the Fogg Art Museum at Harvard; and "Philosophy of the Ostwald System," an article for the *Journal of the Optical Society of America,* July, 1944.

In 1942 Egbert Jacobson, working as art director for the Container Corporation of America in Chicago, and with the generous support of his employer, Walter P. Paepcke, created *The Color Harmony Manual.* (Associated in the project were Carl E. Foss and Walter C. Granville.) This magnificent portfolio, which has since gone into further editions, contains over 900 removable color chips and beautifully fulfills the technical requirements of Ostwald's system. It still ranks as one of the most impressive collections of standards ever issued.

Finally, the best exposition of the overall contribution of Wilhelm Ostwald will be found in Egbert Jacobson's *Basic Color,* published by Paul Theobald at Chicago, 1948. This handsome volume, $8\frac{1}{2}$ x 11 inches, 208 pages, is elaborately illustrated in full color. Emphasis is on esthetics and color harmony.

It is true, of course, that Ostwald's reputation is now firmly established; virtually every book on color published today in the English language pays tribute to him.

Of Ostwald's German publications, the more important ones include: *Die Farbenlehre; Die Harmonie der Farben; Die Harmonie der Formen; Die Welt der Formen; Goethe, Schopenhauer und die Farbenlehre.*

COLOR PLATE I

The color charts on the two following pages are discussed in the Second Chapter, devoted to The Chromatic Colors. Brief comments are included here so that the reader may study the Color Plates by themselves and grasp their significance without having to refer to the text section.

Ostwald remarks that a given color can be changed in four different ways (see Figure 12, page 31). In series 37 to 40, Color Plate I, there is a change in *hue*, in this case from orange to red. In series 41 to 44 there is a change toward white, using purple as an example. In series 45 to 48 there is a change toward black, the scale running from orange to brown. In series 49 to 52 there is a change toward both black and white — or gray — using seagreen. Modifications like these are virtually endless. However, in the Ostwald system they are all subject to measurement and definition.

In series 53 to 60 is what Ostwald terms a random assortment of pure or full colors. Because the eye has a natural sense of order and sequence, most persons would rearrange the random assortment as in the series 61 to 68. This becomes a spectrum and can be readily converted into an effective color circle — such as on Color Plate II.

COLOR PLATE I

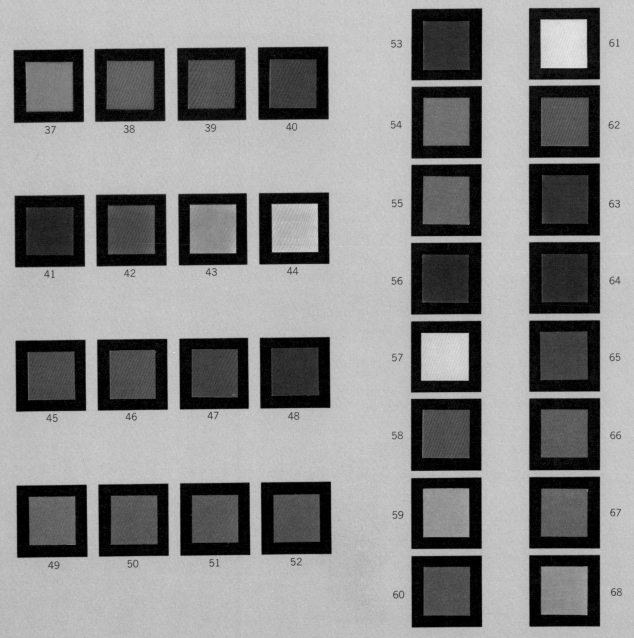

82

COLOR PLATE II

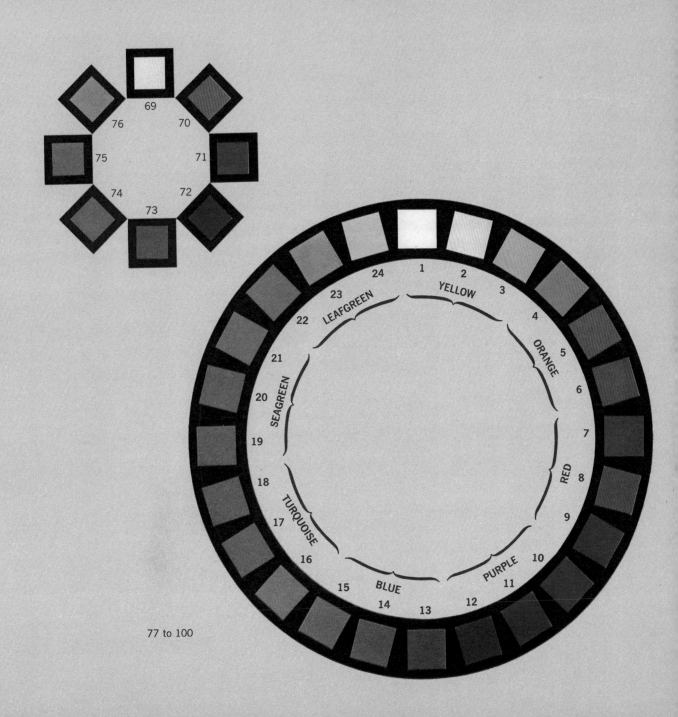

77 to 100

COLOR PLATE II

Being influenced by the great German psychologist Ewald Hering in the design of his color circle, Ostwald chose four fundamental colors, yellow, red, blue, seagreen, not three (red, yellow, blue) as in the past. To the four fundamental colors he added four more, orange, purple, turquoise, leafgreen. Thus there were eight principal colors as shown in series 69 to 76 at the top of Color Plate II, preceding page.

He then proceeded to add intermediate colors until he formed a complete color circle of 24 hues. These he numbered from 1 to 24 starting at yellow.

In the Ostwald color circle the principle colors are yellow 2, orange 5, red 8, purple 11, blue 14, turquoise 17, seagreen 20 and leafgreen 23. These hues and their intermediates are arranged so that perfect visual complements lie opposite each other, yellow and blue, orange and turquoise, red and seagreen, purple and leafgreen, etc. Such pairs will cancel into gray if mixed visually, as on a color wheel (see Figure 13, page 34).

The number 24 is divisible by 2, 3, 4, 6, 8, 12. Where full colors are to be combined harmoniously, these intervals could be used with excellent assurance of beauty.

COLOR PLATE III

The color charts on Plate III, following page, are discussed in the Third Chapter, devoted to Light Clear and Dark Clear Colors. The methods by which Ostwald arrived at neat visual sequences or intervals in color mixture were both unique and ingenious and involved geometrical rather than mathematical progressions. His contribution here — described in the text — is one of the most remarkable in all the long history of color organization.

In series 101 to 107 and 108 to 114 are scales of light colors for red 8 and seagreen 20. Both scale from pa (full color) to na, la, ia, ga, ea, ca to white (see Figure 29, page 55), and the colors contain nothing but white additions.

In series 128 to 132 and 133 to 138 are scales of dark clear colors for red 7 and yellow 3 in which only black has been added to full color.

Ostwald noted that *visual* mixtures of color with white or black do not have the same result as *pigment* mixtures. Series 115 to 118 represents pigment mixtures of blue and white in which the pale tint appears decidedly turquoise. In fact, the shift *with blue pigments* may be as much as one or two steps on the color circle, series 119 to 122.

In a similar fashion, both pigment and visual mixtures of yellow with black tend to produce greenish shades, as in series 133 to 138.

In series 123 to 127, an ultramarine blue graded toward white tends to appear redder in the lighter steps.

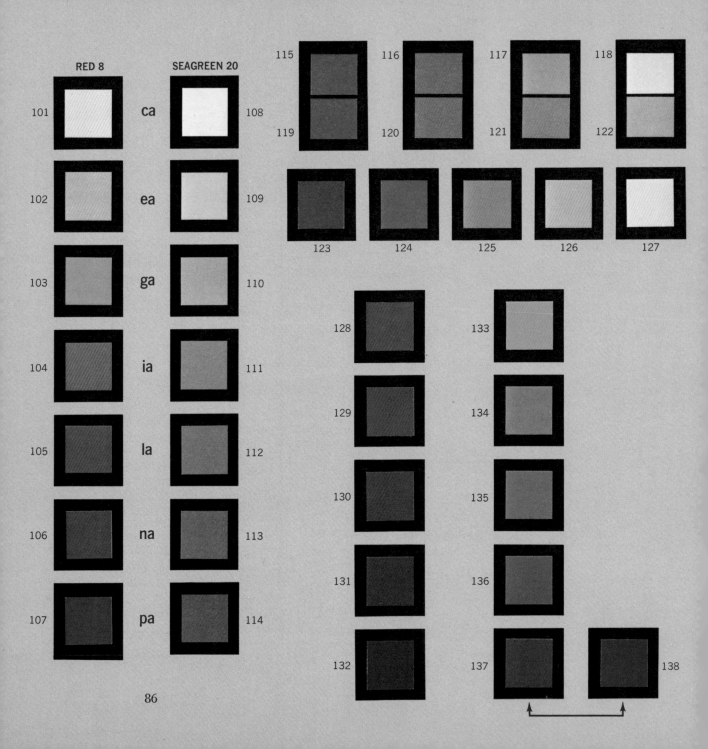

COLOR PLATE III   COLOR PLATE IV

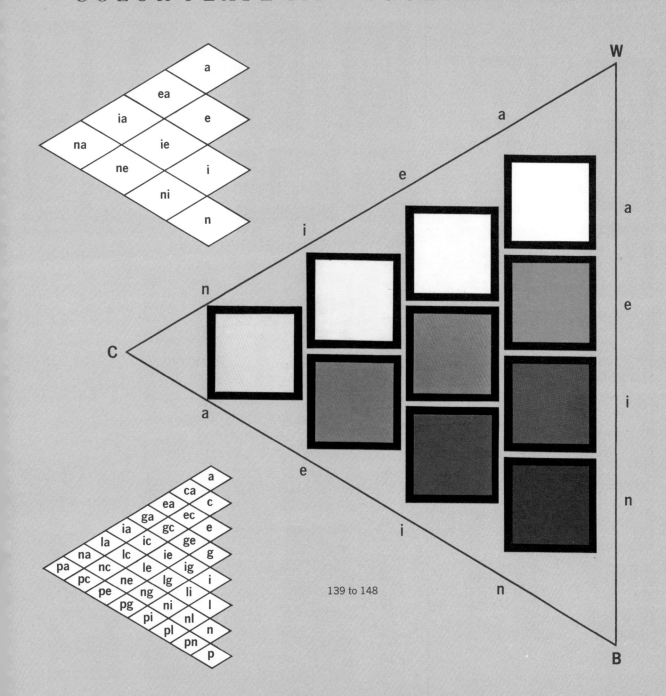

139 to 148

COLOR PLATE IV

The single, most remarkable feature of Ostwald's system is his color triangle, discussed in the Fourth Chapter and shown on the preceding page. The complete notations of the equation are indicated in the lower left black-and-white diagram. For *The Color Primer* Ostwald abbreviated this as in the notations at upper left and as illustrated with yellow in the large triangle. These triangles, one for each of 24 full hues on the color circle, were then arranged like spokes about a wheel to form Ostwald's color solid (Figures 10 and 30, pages 18, 59).

The following points should be observed. The gray scale runs vertically from W to B and has the letters a (for white), c, e, g, i, l, n, p (for black). The light clear colors run parallel to CW. All other scales parallel to CW have equal parts of *black* as on the gray scale (the *second* letter in each notation). The dark clear colors run parallel to CB. All other scales parallel to CB have equal parts of *white* as on the gray scale (the *first* letter in each notation).

The colors that run parallel to WB (ca to pn, ea to pl, ga to pi, ia to pg, la to pe, na to pc) contain apparently equal parts of hue. Ostwald called these scales the shadow series, and he looked upon them as having exceptional beauty. Any variation of any hue, according to Ostwald, necessarily lies within the boundaries of the triangle and could be given an exact notation.

COLOR PLATE V

In Ostwald's *The Color Primer*, Fifth Chapter, four cross sections of an abbreviated solid were illustrated in color. These will be found on Color Plates V and VI, following pages. As in Figure 29, page 55, a complete triangle has 28 steps or notations, plus 8 steps on the gray scale. However, in the abbreviated triangle (see Figure 33, page 61, and Color Plate IV) there are only 6 steps and 4 white, gray, black values.

On Color Plate V are "small color solid" cross sections of yellow 2 and blue 14, and orange 5 and turquoise 17. Color Plate VI shows red 8 and seagreen 20, and purple 11 and leafgreen 23. A key to the notations will be found in Figure 33. Thus Ostwald's 8 principal hues are presented.

As Ostwald explains in the text of his Fifth Chapter, in each principal section the gray scale or "achromatic axis" runs in the center from top to bottom and has the notations a, e, i, n. Colors of equal white content ascend "obliquely upwards," while colors of equal black content descend "obliquely downward." Colors of the shadow series run up and down parallel to the gray scale.

For training purposes in the conception of color organization and to establish sound principles of color harmony, Ostwald has designed a relatively simple instrument of great practical as well as esthetic interest.

COLOR PLATE V

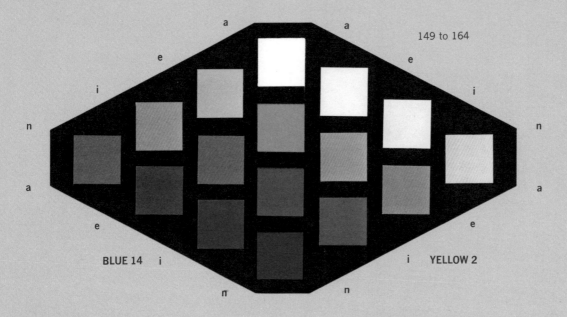

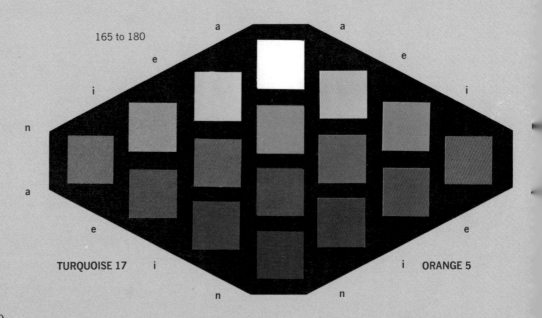

COLOR PLATE VI

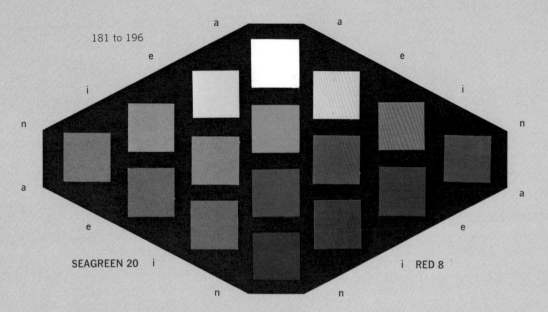

181 to 196

SEAGREEN 20   RED 8

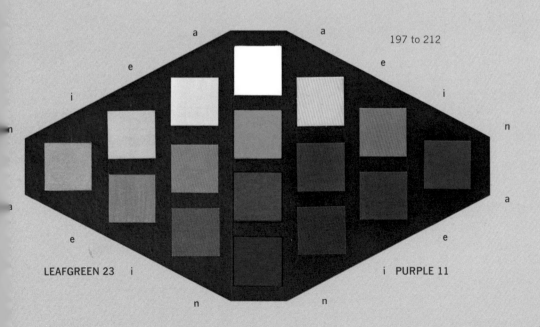

197 to 212

LEAFGREEN 23   PURPLE 11

COLOR PLATE VI

Color harmony in the Ostwald system is founded on the precept that beauty equals order. The colors of a sunset or the rainbow have pleasing order. And this same order, as it may exist in a spectrum of light, Ostwald has brilliantly extended to modified colors in the invention of his color triangle and color solid.

As mentioned, Color Plates V and VI, preceding pages, exhibit cross sections of an abbreviated color solid in which Ostwald's 8 principle hues are arranged as complements. To develop harmonies, the students, the designer, and the artist need but to follow straight paths.

There are the colors of equal black content which run parallel to CW (see Figure 33, page 61). The light clear series are composed of na, ia, ea, and a. Slightly subdued colors are in the series ne, ie, e.

Colors of equal white content run parallel to CB from na to ne, ni, n. Again a softer version of equal-white colors will be found in the series, ia, ie, i.

The shadow series run vertically: ea, ie, ni; and ia, ne.

As to choice of *hue*, this can be based on complements, adjacents, triads, tetrads. In any case, Ostwald insisted that neat intervals should be planned, whether around the color circle, on the triangle, or within the solid itself.

While such principles might not make a great artist, it most certainly would develop a capable colorist.

COLOR PLATE VII

Ostwald took particular pride in his equal value colors and equal value circles — hues having the *same* white and black content. Two examples are shown on the following pages. (Also see Figures 29 and 33, pages 55, 61.) At top are the 8 principal colors at step ea on the color triangle. These are clean pastels, each having the same amount of white in them and no black. At bottom are the same hues at step ni. These are deep shades, each having the same amount of black in them and a small uniform quantity of white — n as on the gray scale.

New to the art of color, up to the time of Ostwald, was this concept that colors having equal white and equal black content were automatically harmonious, for they had orderly and uniform visual balance. Also new was the observation that backgrounds in white, gray, or black were not necessarily neutral and therefore could not be used promiscuously in designs or other works of art. A white ground required the use of colors scaling from pa to a (see Figure 29). The black ground needed colors scaling from pa to p.

As to gray grounds, equal value colors at ne, for example, looked best on n or e grounds. And this held true for any modified or muted series of equal value colors.

COLOR PLATE VII

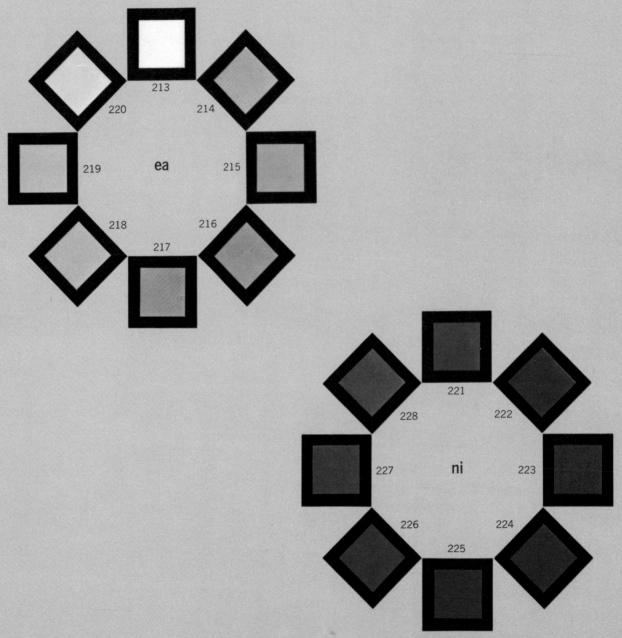

94

COLOR PLATE VIII

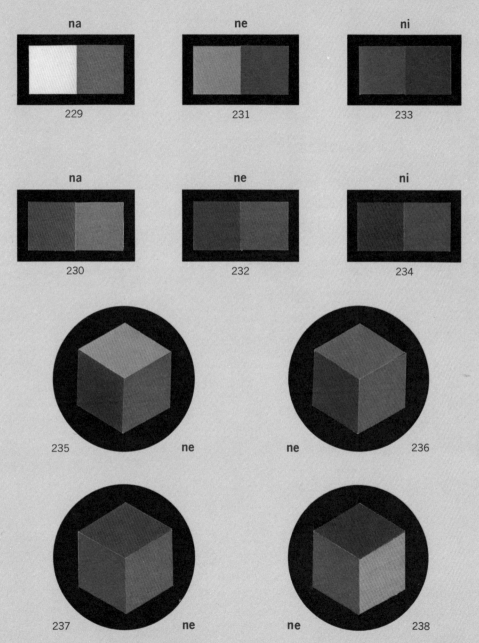

COLOR PLATE VIII

The colors presented on the preceding page are described in Ostwald's Sixth Chapter, "The Harmony of Color." He held a high regard for complementary pairs. In examples 229, 231, 233 were na, ne and ni values for yellow 2 and blue 14. The same values were shown for red 8 and seagreen 20 in examples 230, 232, 234. Effects became softer with decreasing purity. Complementary pairs had an interval of 12 on the color circle (Figure 17, page 38, and Color Plate II). Other hue intervals of 3, 4, 6 were likewise harmonious.

An interval of 8 led to triads. Four such combinations are shown — and all have values at ne. Ostwald remarked that triads in the past had been wrongly defined; the true ones were as exhibited.

In example 235 is a triad (at value ne) of yellow 2, purple 10, and turquoise 18. Example 236 combines orange 4, purple 12, and seagreen 20. Example 237 combines orange 6, blue 14, and leafgreen 22. Example 238 combines red 8, turquoise 16, and leafgreen 24. To these equal white-and-black-content harmonies, suitable achromatic colors to go with them would be gray n or e.

In effect, the Ostwald system became a useful tool that established color relationships in terms of human vision. For this reason any exploration of the solid in any direction led to beauty.